INVASIVE PYTHONS in the United States

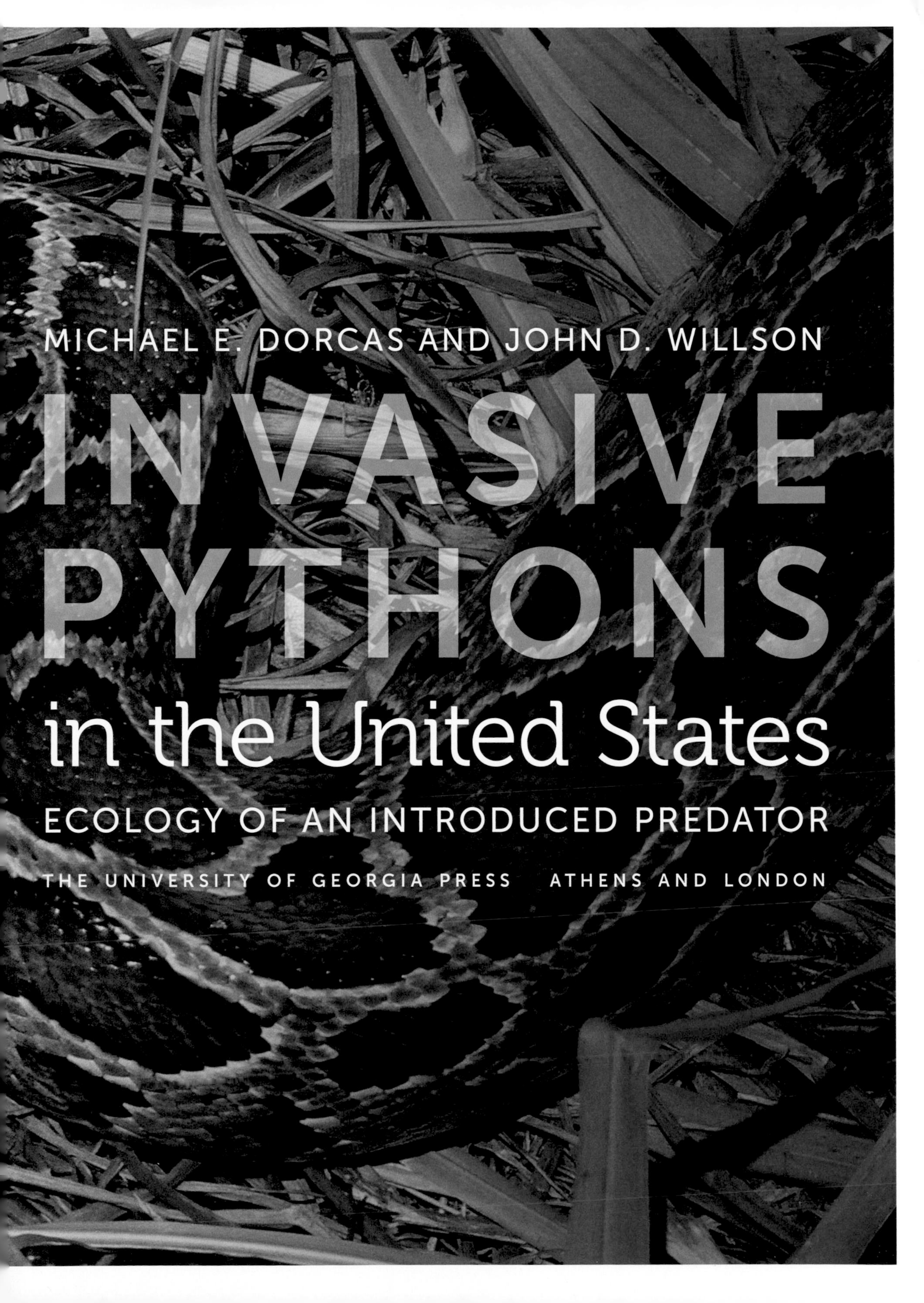

MICHAEL E. DORCAS AND JOHN D. WILLSON

INVASIVE PYTHONS in the United States

ECOLOGY OF AN INTRODUCED PREDATOR

THE UNIVERSITY OF GEORGIA PRESS ATHENS AND LONDON

Athens, Georgia 30602
www.ugapress.org

Designed by Mindy Basinger Hill
Set in Minion Pro
Printed and bound by Four Colour Print Group
The paper in this book meets the guidelines for permanence and durability of the Committee on Production Guidelines for Book Longevity of the Council on Library Resources.

Printed in China
15 14 13 12 11 P 5 4 3 2 1

Library of Congress Cataloging-in-Publication Data
Dorcas, Michael E., 1963–
Invasive pythons in the United States : ecology of an introduced predator / Michael E. Dorcas and John D. Willson.
p. cm.
Includes index.
"A Wormsloe Foundation Nature Book."
ISBN-13: 978-0-8203-3835-4 (pbk. : alk. paper)
ISBN-10: 0-8203-3835-4 (pbk. : alk. paper)
1. Burmese python—Florida. 2. Python (Genus)—United States. 3. Introduced reptiles—United States. I. Willson, John D., 1979– II. Title.
QL666.067D667 2011
597.96'781718097599—dc23 2011012850

British Library Cataloging-in-Publication Data available

For his tireless efforts to understand invasive pythons and the people who are fascinated by them, we dedicate this book to Skip Snow.

Contents

Foreword

The word "python" conjures up the image of a snake of enormous size and powerful constriction, a stealthy, well-camouflaged predator of deliberate and unhurried movement. American readers and moviegoers are familiar with pythons through the tales of Edgar Rice Burroughs' *Tarzan in Africa* and Rudyard Kipling's *The Jungle Book* in India.

Now nonfiction stories and real-life videos about pythons abound. From daily newspapers and popular magazines to TV news shows, documentaries, and YouTube, pythons are in the public eye. Why? Because Burmese pythons have invaded the United States.

Introduced species cause endless problems across the world. Some, such as European rabbits in Australia, mongooses in Jamaica, and brown treesnakes in Guam, have caused severe economic damage and have had permanent negative impacts on native wildlife. In the United States, imported fire ants, cogon grass, and Asiatic carp are but a few of numerous unpopular exotic plants and animals that have earned the epithet "invasive species." But the Burmese python's takeover of southern Florida, which was not generally recognized as a genuine problem until the 21st century, is unprecedented. The python invasion may rival all others in terms of its potential to completely alter the structure of native ecosystems and to capture the public's attention.

The number of Florida's larger mammals and magnificent birds that are suitable prey for these exotic predators is staggering. Burmese pythons, which begin life as 2-foot-long hatchlings, can reach lengths greater than 20 feet. Almost any warm-blooded animal, small or large, is potential prey for these giant predators, which will even eat alligators. Invasive pythons in the United States have become an environmental specter that urgently warrants public concern.

In *Invasive Pythons in the United States*, Michael Dorcas and John D. Willson describe the biology of the Burmese pythons now naturalized in Everglades National Park and surrounding habitats. They provide a thorough discussion of python natural history with particular emphasis on the Burmese and Indian pythons, which most scientists consider a single species. The species occupies a vast geographic range in Southeast Asia and the Indian subcontinent and inhabits a variety of habitats ranging from tropical jungles to near-deserts and the temperate foothills of the Himalayas. The authors also discuss how big pythons really get, their food and feeding habits, their reproduction, and their natural predators.

Understanding python biology in the snake's native range is essential for assessing the adaptive potential of these reptiles to invade their new ecosystem in southern Florida. The authors review current knowledge regarding the distribution of pythons in Florida, how they became established there, and the probable size of the current population. They also speculate on what the future holds for these giant snakes. How likely are they to expand their range up the Florida peninsula—possibly to other southern states? How fast are python populations likely to grow in Florida or elsewhere under various climatic conditions and with varying food resources is a key consideration in determining their potential impact.

Of particular relevance is the effect Burmese pythons have already had on native wildlife and the risks they pose to domestic animals and pets, including chickens, dogs, cats, sheep, and goats. An extremely sensitive aspect is the risk to humans from one of the largest snakes in the world. Records exist of pythons eating people in the wild, and the tragedy of children being killed by pet pythons in the United States is much too real. Would a large python attack a child or even an adult alongside a lagoon in southern Florida if given the opportunity? The odds are vanishingly small, but people's perceptions often drive public sentiment about such issues. As always, learning the facts is the best antidote for unwarranted fears. This book provides such critical information.

Finally, the authors address the risk from other species of large constrictors, such as rock pythons and boa constrictors, which are already established in the subtropical habitats of southern Florida, albeit on a smaller scale than Burmese pythons, as well as anacondas from tropical America and reticulated pythons from Asia.

Stories about invasions of giant predators from other lands were once limited to Hollywood horror movies and science fiction pulp magazines. Today those stories are the new reality. *Invasive Pythons in the United States* sums up the current scientific facts that can be brought to bear on this increasingly serious problem.

Whit Gibbons

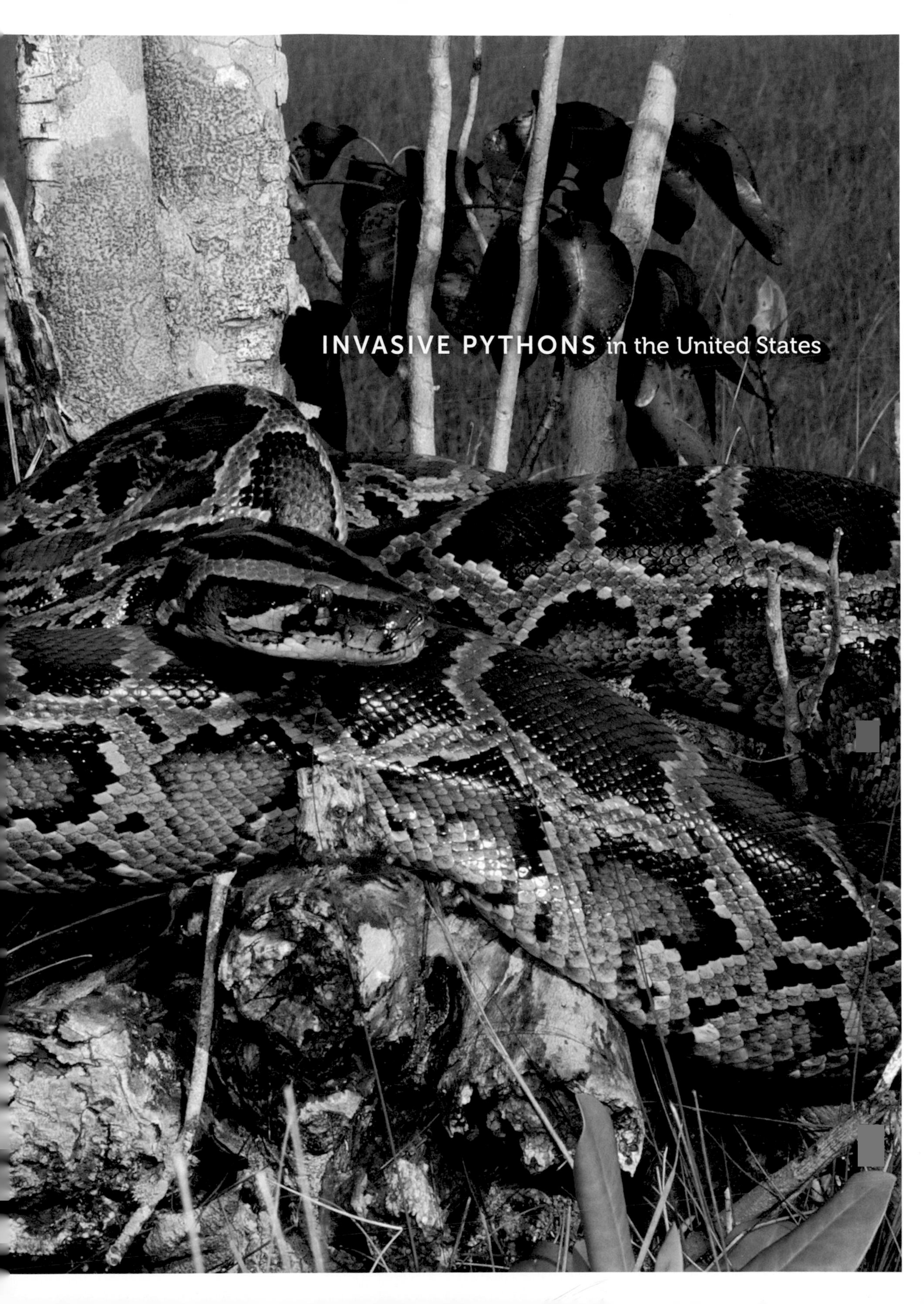

INVASIVE PYTHONS in the United States

Introduction

Nearly everyone is fascinated by giant snakes, even people who are afraid of them. But pythons right here in the United States? How are we to respond to that? Most people think of pythons as giant snakes that live in distant tropical jungles. But one of the world's largest snakes is now thriving in South Florida. The United States is home to about 130 species of native snakes ranging in size from the diminutive blind snake, which can be smaller than an earthworm, to the indigo snake, which can attain a length greater than 8 feet. But even the largest indigo snakes do not approach the size attained by the Burmese pythons now living in South Florida, which can reach lengths of more than 20 feet and weigh hundreds of pounds.

Since 2001, over 1,600 Burmese pythons have been captured in Everglades National Park and surrounding areas.

Exactly how Burmese pythons (*Python molurus bivittatus*) were introduced to South Florida is unknown. These natives of Southeast Asia are commonly kept as pets, and over the years large numbers have been imported into the United States and bred in captivity. Most authorities agree that the original source of the pythons now present in South Florida is the reptile pet trade. Whether the snakes were released intentionally, accidentally, or both may never be known.

We do know that pythons have been common, breeding, and widespread throughout hundreds of square miles of southern Florida since the turn of the 21st century. The python population appears to be increasing and expanding outward from Everglades National Park northward through the Florida peninsula and southward into the Florida Keys. In fact, pythons are becoming relatively common in many parts of South Florida. Nobody knows exactly how many of them are now living in Florida, but rough population estimates range from tens of thousands to hundreds of thousands. Perhaps even more disturbing, Burmese pythons are known to inhabit temperate regions in Asia, and research suggests that suitable climates and habitats exist throughout much of the southern United States. Exactly how far pythons will be able to spread will be a major factor determining their ultimate impact on our environment. Pythons affect native ecosystems

Burmese pythons more than 16 feet long have been found in the Everglades.

Burmese pythons are regularly found basking along canal levees across much of South Florida.

This photograph of a Burmese python that was found dead in 2005 in a remote area of the Everglades with an American alligator protruding from its stomach made headlines around the United States.

Because of their secretive behavior and excellent camouflage, pythons are virtually invisible except when crossing roads.

by eating other animals, primarily mammals and large birds, and are known to prey on everything from herons to bobcats to white-tailed deer. The exact impact of python predation on native wildlife remains unknown, however, and is currently the subject of scientific investigation. If mammal and bird populations should decrease because of the pythons, the effects of such population declines will potentially have major impacts on entire ecosystems.

Of paramount concern to the U.S. public is the potential danger these large snakes pose to people. In general, snakes have an undeserved bad reputation. Like most other animals, snakes are generally afraid of people. Snakes that appear "aggressive" by trying to bite are actually acting in self-defense, trying to dissuade an attacker from hurting them. Think about it from the snake's perspective: you might try to bite too if an animal much larger than yourself was trying to hurt you. Although on rare occasions a large python might view a human as a potential meal, even large pythons typically respond defensively toward people and would rather be left alone. Nevertheless, large pythons have killed and eaten people in Asia, and captive Burmese pythons have killed people in the United States. As the python population increases and expands its range into more densely populated areas, the likelihood of human-python encounters increases, and deaths resulting from wild pythons must be considered a possibility.

Pythons in South Florida have been known to eat white-tailed deer.

PURPOSE OF THE BOOK

The extensive media coverage of invasive Burmese pythons over the last few years has generated considerable interest in this issue from both the general public and the scientific community. Politicians have spoken out about pythons and the threats, real or imagined, that they pose. Unfortunately, the combination of "snake phobia" and media sensationalism has resulted in many misconceptions about the realities of the python situation in southern Florida. To date, published information has been restricted to a few scientific publications and a vast amount of popular press coverage (such as Internet, newspapers, and television) that varies dramatically in authenticity and accuracy.

Our primary goals in this book are to provide accurate information about invasive Burmese pythons in the United States and to address concerns about their potential impacts. We include information about general python biology, the biology of Burmese pythons in their native range, research on pythons in the United States, the history and status of the introduced pythons in southern Florida, the risks pythons pose in Florida and elsewhere, potential methods to control python populations, pythons in the pet and skin trades, and other boa and python species that are established or are perceived to be at risk of becoming established in the United States. The articles and books we mention are listed in the Further Reading section at the back of the book along with other sources of information on pythons. We hope this work provides an informative resource for both scientists and the public that will help to clarify issues related to these invasive snakes and quell the hysteria associated with this problem.

Pythons have been found with increasing regularity in Everglades National Park since the late 1990s.

INVASIVE PLANTS AND ANIMALS

Nonnative plants and animals present an enormous threat to biodiversity in the world today. In the United States alone, the effects and management of invasive species cost more than $100 billion annually.

The invasive vine kudzu has overrun large areas of forest in the southeastern United States.

The distinction between the terms "introduced," "nonnative," "exotic," and "invasive" is not always clear, and they are sometimes used interchangeably. Most scientists consider invasive plants and animals to be those that have measurable impacts on the ecosystems into which they have been introduced or those that have caused measurable economic damage. Nearly 50 percent of the imperiled native species in the United States are endangered at least partly because of invasive species.

Invasive species can affect their environment in numerous ways. Some can outcompete native species, leading to decline or extinction of native animals or plants. Many invasive plants alter the physical structure of the environment, creating conditions inhospitable for native species. Others are toxic to native animals, which often do not know to avoid eating them. Some invasive animal species negatively affect the environment by preying directly on native species. Many scientists believe that only habitat destruction poses a greater threat to biodiversity than invasive species do.

Among the most notorious invasive plants in the southeastern United States is kudzu (*Pueraria lobata*), a vine imported from Japan that was planted throughout the Southeast in the early 1900s to prevent soil erosion. The climate in the Southeast is perfect for kudzu, and it has become a major problem in many areas. Kudzu grows rapidly and literally covers natural vegetation, eventually killing it. Annual costs resulting from crop loss and for kudzu control approach $500 million. The red imported fire ant (*Solenopsis invicta*) was accidentally introduced in Mobile, Alabama, in the 1930s and is now found in disturbed areas throughout the South. Fire ants represent a major threat to many animals. The U.S. Food and Drug Administration estimates that fire ant damage and control cost more than $5 billion per year. Cane toads (*Rhinella marina*), native to Central and South America, have been introduced into many areas, including southern Florida, Hawaii, and Australia, where they were intentionally introduced to control cane beetles. These very large toads have glands on their neck and back that produce a potent poison. In Australia large native lizards, birds, mammals, and snakes eat the toads and are killed by their toxins. Populations of many native predators have declined as a result of the cane toad invasion. Ironically, the cane toads had little effect on the cane beetles they were introduced to control.

Invasive plants such as kudzu and water hyacinth can drastically alter natural ecosystems.

Some introduced exotic species are apparently not invasive; that is, they seem to have little or no measurable negative impact on natural ecosystems. In Florida and elsewhere, the Mediterranean house gecko (*Hemidactylus turcicus*) is now common around houses and other buildings. The geckos are frequently found near lights at night feeding on insects, but apparently do not have any major impact on other native species. Likewise, the tiny Brahminy blind snake (*Ramphotyphlops braminus*) is an Old World native that has been introduced into Florida through the ornamental plant business. This little earthworm-like snake is often found under logs and rocks feeding on ants and termites, but seems to have no appreciable impact on natural ecosystems.

WHY INVASIVE PYTHONS ARE IMPORTANT

Introduced Burmese pythons are an important concern for several reasons. First, they are top predators. The largest pythons are capable of eating nearly any native bird or mammal in the southeastern United States; some can even eat medium-sized American alligators. None of South Florida's native animals evolved in the presence of a giant

The smallest snake currently found in the United States is the introduced Brahminy blind snake, which is sometimes called the flowerpot snake.

The cane toad, introduced in many areas outside its native range in the American tropics, is among the most devastating invasive species. In Australia cane toads have caused declines in many native animals.

Introduced Mediterranean geckos are common in many areas of the southern United States. Because they are generally restricted to urban areas, they probably pose little threat to native wildlife.

Pythons consume a wide variety of birds and mammals. This python (left) was found while constricting a white ibis. Wading birds such as this wood stork (right) are icons of the Everglades. The impact invasive pythons will have on these species remains unknown.

constricting snake. Consequently, they may be more vulnerable to predation because they may not perceive pythons as dangerous or may not be adept at avoiding them. Further, the birds and mammals that pythons eat play important roles in the food chain. Such impacts on key species in food chains can alter entire ecosystems, often in unpredictable ways.

Burmese pythons are also important because they threaten imperiled species. Numerous species of threatened and endangered birds and mammals inhabit the unique ecosystems of southern Florida, and pythons have already been documented preying on some of them. As the python population expands, such species of high conservation concern could be severely affected or even extirpated by this new predator prowling their environment.

The potential danger that Burmese pythons pose to domestic wildlife and possibly even to humans is another reason to be concerned about their presence. Pythons have already been documented feeding on some domestic animals, and as their populations expand into areas where humans live, encounters with dogs, cats, chickens, and other domestic species are likely to increase. Of course, pythons may also help to control rat populations in agricultural areas, thus providing an unexpected benefit to some forms of agriculture. The danger pythons pose to humans is small but nonetheless real. Pet Burmese pythons do sometimes kill humans. As of 2010, no fatal attacks on humans by a wild Burmese python have been reported in Florida, but the possibility of such encounters will certainly increase.

Although invasive pythons are currently limited to southern Florida, recent research indicates that they may be able to expand their range outside that area. Some climate models suggest that much of the southern United States may provide suitable conditions for pythons, and the problems associated with pythons in South Florida could thus eventually become much more widespread. For all these reasons, the Burmese python should be considered a serious and harmful invasive species.

1 About Pythons

Like all snakes, pythons are reptiles, members of a group that traditionally has included turtles, lizards, crocodilians, and the rare lizard-like tuatara. Although recent studies have shown that lizards and snakes as a group are only distantly related to turtles and crocodilians, the latter of which are more closely related to birds, most laypeople and many scientists continue to use the traditional classification. Like all reptiles, birds, and mammals, snakes are amniotes; that is, either they produce eggs adapted for being laid on land, not in the water, or they give birth to live young. Also like all reptiles, snakes have scales covering the entire body. The characteristic that makes snakes distinctive from most other reptiles is their long, limbless body. Although they have no limbs, snakes have some abilities that few other animals can rival. Many are well adapted for living underground, in water, or even in trees.

Snakes use their forked tongue to detect chemicals in the air.

Collectively, snakes and lizards make up a group of reptiles known as squamates. Among other characteristics that they have in common, squamate males have two penises (known as *hemipenes*) located in the base of their tail. They use only one at a time for mating, but they can use either their right or left hemipenis. Many species of lizards resemble snakes in being elongated and lacking legs, but most lizards have movable eyelids and external ear openings, which snakes lack. Finally, snakes cannot regenerate their tail the way most species of lizards can.

The scales that cover the body of snakes and protect them from dehydration and injury are made of thickened, folded epidermis—the outer layer of skin. Periodically, snakes undergo a process called *ecdysis* in which they shed, or molt, the outermost layer of their skin, including the transparent scale that covers and protects each eye. They often shed the skin in one piece, turning the old skin inside out as they rub it off their body, much like a person pulling off a sock.

Most snakes have many sharp teeth that curve backward toward their throat. Venomous species usually have enlarged hollow teeth called *fangs* that can inject venom into prey or potential predators. Because their teeth are generally designed for grabbing and holding

Legless lizards (top left) differ from snakes in having ear openings and movable eyelids. Snakes (bottom left) lack ear openings and have clear scales covering their eyes. Their flexible jaws allow snakes (right) to eat very large prey relative to their body size.

on to prey, rather than for chewing, virtually all snakes swallow their prey whole. Their jaws are loosely connected to their skull, allowing snakes to stretch their mouth over their prey and consume very large animals, sometimes even larger than themselves. In fact, some snakes have consumed prey weighing more than 150 percent of their own body weight. All snakes eat other animals, although a few have been observed eating fruit, presumably by mistake. Some snakes, such as North American racers (*Coluber constrictor*) and cottonmouths (*Agkistrodon piscivorus*), are generalists that eat a wide variety of animal species, including insects, fish, mammals, birds, amphibians, lizards, turtles, and even other snakes. Others have extremely specialized diets, such as the several species of crayfish snakes (genus *Regina*) that eat almost exclusively crayfish as adults. Snakes that do not use venom to subdue their prey either overpower the animal and swallow it alive or kill it first by constriction. Constrictors typically grab their prey with their mouth and then quickly wrap around and squeeze the prey animal until its breathing or heart stops.

No snake has movable eyelids or external ear openings.

Internally, snakes have the same basic organs and organ systems as other vertebrates. Snakes have between 200 and 400 vertebrae, most of which have a pair of ribs attached to them. The muscles that move the body attach to the vertebrae and ribs. Because snakes are elongated, paired organs such as kidneys are often positioned one behind the other rather than side-by-side, and single organs such as the liver are elongated.

The cottonmouth (top) eats a variety of prey including fish, birds, mammals, insects, frogs, salamanders, lizards, turtles, and snakes, even other cottonmouths. The eastern hognose snake (bottom) is a dietary specialist that feeds almost exclusively on toads.

Pythons have small "spurs" on either side of the cloaca that are actually the tiny remains of hind limbs. Male pythons have larger spurs than females and use them to stroke females during courtship.

Although they are not venomous, pythons have large, recurved teeth that allow them to keep a strong grip on prey.

Many snakes have only one functional lung, although boas and pythons have two. The snake's powerful digestive system can typically digest every part of the prey, including bones, although hair and feathers generally pass through the digestive tract intact.

The large intestine, reproductive tract, and urinary tract all empty through a single opening called the *cloaca* that demarcates the snake's body and tail. When mating, the male snake inserts one of his two hemipenes into the female's cloaca. Once inside the female, the hemipenis fills with blood and enlarges, and its numerous hooks and spines anchor it in place. The two mating snakes may remain attached in this way for several hours while the male's sperm travels along a groove on the hemipenis to the female snake's reproductive tract. Some female snakes can store live sperm for long periods, possibly even years.

Although snakes do not have external ear openings, they have the same basic senses that humans and most other animals have. They have a very well developed chemosensory system involving their forked tongue and small organs in the roof of the mouth called *Jacobson's organs*. The tips of the tongue collect particles from the air or from surfaces and place them in the Jacobson's organs, where chemoreceptive cells help the snake to determine characteristics of its environment, much as if the snake were tasting the air rather than smelling it.

Like all reptiles, snakes are cold-blooded; that is, a snake's body temperature is determined by a combination of environmental conditions and its behavior. Scientists use the term "ectothermy" rather than "cold-blooded" to describe this phenomenon, and thus a reptile is considered an "ectotherm." Although they are often considered more primitive than the warm-blooded, or "endothermic," birds and mammals, ectotherms have many advantages over endotherms. Because ectotherms do not have to use metabolic energy to maintain a high, stable body temperature as birds and mammals do, they require much less food. This low-energy lifestyle allows many reptiles, and especially snakes, to survive long periods without food. It is not uncommon for snakes to remain healthy after going an entire year without eating. When food is plentiful, snakes can devote any extra energy into growth and high reproductive output, and some species grow very rapidly under favorable conditions.

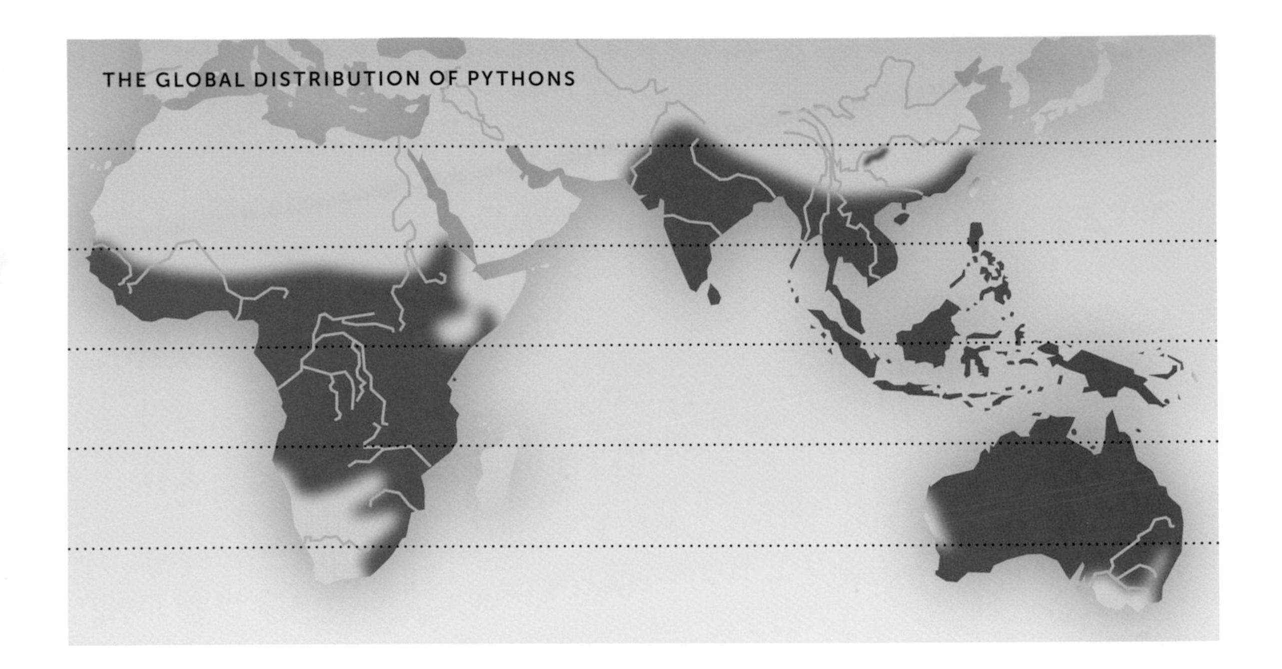

PYTHON BIOLOGY

Pythons are classified in the family Pythonidae, one of approximately 20 families of snakes worldwide. There are at least eight genera (related groups) of python species and at least 26 species. Pythons are closely related to boas, which belong to the family Boidae; in fact, some scientists believe that boas and pythons are closely related enough to be in the same family. The boas include the familiar boa constrictor (*Boa constrictor*) found throughout most of Latin America and the infamous green anaconda (*Eunectes murinus*) of South America. Pythons are found strictly in the Old World, including Africa, southern Asia, Malaysia, Indonesia, Papua New Guinea, and Australia. All pythons are nonvenomous animals that kill their prey by constriction.

Pythons inhabit a variety of habitats but are native to the Old World.

Although they are generally large snakes, pythons range in size from the relatively small Children's python (*Antaresia childreni*) from Australia, which typically gets no more than 3 feet (91 cm) long, to the giant reticulated python (*Python reticulatus*), which may exceed 30 feet (9.1 m) in length. The family includes five of the six largest snake species in the world.

Pythons and boas are considered to be primitive snakes—more closely resembling the lizard-like ancestor from which they evolved—because they retain remnants of a pelvis and have tiny vestigial limbs known as *spurs*. The latter can be seen on either side of the python's cloaca (at the base of the tail) as small clawlike structures. The spurs are larger

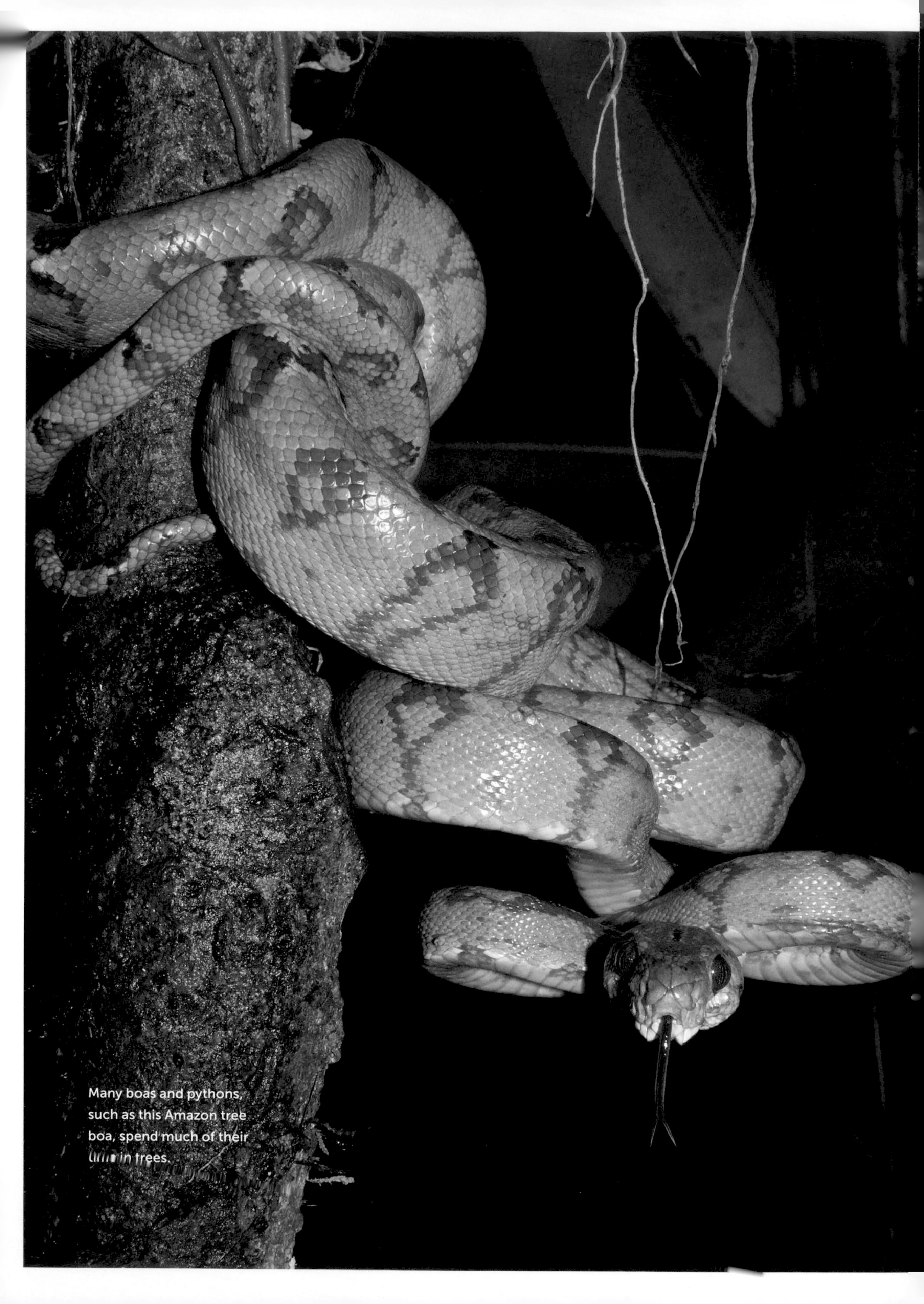

Many boas and pythons, such as this Amazon tree boa, spend much of their [illegible] in trees.

Not all boas and pythons are large tropical animals. The secretive rubber boa is only 2 feet long and is found throughout the northwestern United States and even in southern Canada.

in males, which use their spurs to stroke the female during courtship. Most pythons and many boas have infrared receptors on their lips similar to the heat-sensing pits of pitvipers. In some cases these receptors can detect differences in temperature as small as a fraction of a degree, allowing the snake to sense prey that differs in temperature from the surroundings (such as warm-blooded mammals and birds). The sensitivity of the receptors at different distances from the snake is unknown, but they are apparently somewhat less sensitive than those of pitvipers. Some evidence indicates that snakes sometimes use their infrared receptors for thermoregulatory purposes (e.g., finding a warm spot in which to bask) or for detecting predators.

All pythons are oviparous, meaning that they lay eggs. Unlike most snakes, which deposit their eggs and then leave, many female pythons coil around their eggs to

Like most pythons, this green tree python has infrared receptors on its lips that allow it to sense prey whose temperature differs from the surroundings.

protect them until they hatch. Females of some species contract their muscles spasmodically while they are coiled around their eggs in a process known as *shivering thermogenesis*. The muscle contractions produce heat that warms the eggs and speeds up their incubation. Because of this increase in her body temperature (and thus the eggs' temperature as well) the female python actually becomes warm-blooded, or endothermic, when she is gestating her eggs. The protection and warmth the female python provides to her developing eggs likely increase their developmental rate and chances of survival.

Burmese pythons' beautiful coloration, impressive size, and hardiness in captivity have made them favorites among reptile hobbyists.

Pythons occupy a variety of habitats in their native range, including deserts, tropical rainforests, grasslands, and even urban environments. Some species, such as the green tree python (*Morelia viridis*), are almost always found in trees, where they feed primarily on birds; others spend considerable amounts of time underground or at least under vegetation or piles of debris. The largest pythons often spend substantial amounts of time in aquatic habitats, where they ambush prey at the edge of the water. Additionally, spending time in water, where they are buoyant, may make it easier for large pythons to move from place to place.

PYTHONS AS PETS

Pythons have long been popular pets, and several species make excellent pets for people who are prepared to care for them. Many pythons become tame in captivity and are reluctant to bite defensively, and some species are relatively easy to maintain. Many reptile hobbyists favor the ball python (*Python regius*) from West Africa, a stocky but relatively small snake that often tames down well in captivity and has minimal special husbandry requirements. Burmese and reticulated pythons (*P. reticulatus*) are bred in captivity in huge numbers each year for the pet trade, and some species, such as ball pythons, are imported from the

Many herpetologists got their start by owning a boa or python. Children should handle large constrictors only when supervised by an adult experienced with keeping large snakes.

The hundreds of miles of human-made canals that crisscross South Florida provide ideal habitat for invasive pythons.

wild by the thousands as well. Many herpetologists, including the authors of this book, became interested in herpetology by keeping snakes, including boas and pythons, as pets. Pythons and other snakes are also excellent educational tools. Few props can captivate an audience more effectively than a large python. The role that pythons have played in inspiring young scientists and promoting wildlife appreciation and conservation should never be overlooked.

Any pet, but particularly a large constrictor, incurs a significant responsibility on the part of the owner. Pythons can live for more than 25 years, but many owners tire of taking care of their large pets well before then, resulting in neglect or poor husbandry. Finding someone to purchase or adopt a large python may be quite difficult. Most zoos do not want more than a few large pythons, and many people who do want to keep pythons want to purchase babies and raise them to adulthood themselves. Unfortunately, some pythons escape or are intentionally released when owners do not want them anymore. Although most of these snakes die or are killed by humans, release or escape of pet pythons likely resulted in the python and boa populations that are now established and spreading in South Florida.

PYTHONS AS POTENTIALLY INVASIVE SPECIES

Certain species of animals and plants have characteristics that make them particularly prone to survive outside their native habitat. Many of the species that are commonly exported for the ornamental plant and pet trades are considered desirable partly because they are hardy and easy to propagate. Species that cling to objects, including many insects, treefrogs, and lizards, and those that are particularly good at remaining hidden are often unintentionally transported to new areas as stowaways and end up as introduced species. Whether such accidentally introduced species become established or spread invasively depends on other characteristics of the species and on the area of introduction.

Female Burmese pythons remain with their eggs until they hatch.

Invasive snakes can devastate native ecosystems. In the 1950s, for example, the brown treesnake, a mid-sized climbing snake native to Australia and New Guinea, was accidentally introduced to the U.S. territory of Guam in the north Pacific. Treesnake populations exploded over the next few decades, and soon nearly every one of the island's native bird and bat species had been driven to extinction. Treesnakes remain abundant on Guam, where virtually the only other vertebrates present now are small lizards and introduced rats. Years of research on methods to control brown treesnakes on Guam have been somewhat successful, but complete eradication seems unlikely.

When species are imported and bred in large numbers, as is the case with pythons in the United States, the potential for accidental escape or intentional release is relatively high. In fact, released or escaped pythons of various species are regularly found throughout the United States. In the vast majority of cases these introduced snakes do not establish a reproducing population—either because the habitat or climate is unsuitable for them, because predators or humans kill them, or because they cannot find a mate or sufficient food. Pythons are generally thought of as tropical animals unlikely to become established in nontropical regions. Before it became apparent that Burmese pythons were living and reproducing in the Everglades, which periodically experiences freezing temperatures, most scientists would have considered establishment of a reproducing population there unlikely or impossible. We can now look retrospectively at characteristics of this species that contributed to its establishment and remarkable population increase in southern Florida.

The impressive reproductive potential of Burmese pythons has been a major factor in the establishment of this species in Florida. A typical adult female python in Everglades National Park reproduces about every 2 years and produces an average of 40 offspring. The fact that the female incubates and protects her eggs increases the likelihood of successful hatching. And although a variety of animals can and likely do eat hatchling pythons, the hatchlings are too large (18–36 inches [45–90 cm]) to be prey for many of the native predators that typically eat young

A high reproductive rate is one of the traits that make pythons successful invaders.

snakes. If food is plentiful, young Burmese pythons can grow at remarkable rates—up to 8 inches (20 cm) per month. Growth rates slow as the snakes become older, but Burmese pythons can attain reproductive maturity within 3 to 4 years of hatching.

Finding mates is often a limiting factor in the establishment of introduced species. That is, if a single female is introduced and cannot find an introduced male, she may live for many years but be unable to reproduce in her new environment. Two characteristics of Burmese pythons may have helped to overcome that limitation. First, female Burmese pythons may be able to store viable sperm for long periods—perhaps even years—before they use it to fertilize their eggs. Thus, if a female python mated in captivity and then escaped or was released, she might be able to produce viable offspring even in the absence of a mate. Second, there is evidence that female Burmese pythons can produce viable offspring without mating through a process called *facultative parthenogenesis*. In a well-documented case, Thomas Groot from the University of Amsterdam and his colleagues used

Although invasive pythons have few predators, large American alligators have been observed eating pythons on several occasions.

molecular techniques to prove that a female Burmese python produced viable offspring without mating, and these offspring were genetically identical with their mother. Parthenogenesis has been documented, albeit rarely, in other species of snakes as well. The Brahminy blind snake (*Ramphotyphlops braminus*), a species native to Asia, is entirely parthenogenetic, and its ability to reproduce without mating has undoubtedly contributed to its success as an invader in many locations around the world, including Florida and Hawaii.

Introduced predators that do not have narrow habitat or food requirements tend to be more likely to become established outside their native range. Burmese pythons will apparently eat nearly any bird or mammal they can capture and swallow, regardless of its size. Further, because Burmese pythons are cold-blooded (ectothermic), their overall food requirements are far lower than those of warm-blooded predators. When food is plentiful they can devote substantial amounts of energy to growth and reproduction, and when food is scarce they can survive for very long periods without eating.

Pythons tracked using radiotelemetry moved more than 25 miles within just a couple of months.

Certain characteristics of South Florida may also have contributed substantially to the successful introduction and establishment of invasive Burmese pythons. First, many of the reptile breeders and importers in the United States are based in South Florida, thus increasing the possibility of exotic reptile introductions there. The subtropical climate of South Florida has likewise certainly contributed to the successful establishment of pythons, as has the lack of other extremely large snakes that might compete with pythons. American alligators are probably most similar to pythons in terms of their habitat and diet, but alligators generally eat cold-blooded prey such as fish, snakes, and turtles. Because pythons tend to be aquatic, the many artificial canals and other water bodies present in South Florida increase the number of habitats available to pythons and may facilitate their movements and ability to capture prey.

In retrospect, the combination of the python's characteristics and availability for introduction within the suitable environment of southern Florida created what some might call a "perfect storm," enabling the establishment of a reproducing and expanding population of invasive snakes. Now that they have become established over a large area, pythons are unlikely to be eliminated within the foreseeable future, either by humans or through any natural processes. The impacts they will have and the extent of their spread remain to be seen.

2 Natural History of Indian and Burmese Pythons

Most authorities consider the Burmese python to be a subspecies, or race, of the Indian python, *Python molurus*, and give it the scientific name *P. molurus bivittatus*. Indian pythons are known primarily for their large size. Their maximum size is debated, but they certainly attain lengths greater than 20 feet (6 m) and may, on extremely rare occasions, exceed 25 feet (7.5 m) in total length. Regardless of the maximum length actually attained, adult Indian pythons are massive, heavy-bodied snakes capable of killing and eating most other animals in their environment. Like many other pythons, they have numerous recurved, sharp teeth; vertical pupils; many body scales; and relatively small belly (ventral) scales, or *scutes*, when compared with other snakes. The tail is relatively short, generally constituting approximately 12 percent of the total body length. Color pattern is variable but the back usually has 30 to 40 dark brown, irregular, black-bordered blotches on a lighter background of tan, gray, gold, or yellow. The skin is usually very shiny and sometimes is even iridescent. The sides have irregular blotches and markings, and the head generally has a dark, arrow-shaped mark. The belly is usually white or yellowish.

Indian pythons (*Python molurus molurus*) are often lighter in color than Burmese pythons.

Depending on the authority, either two or three subspecies of the Indian python are recognized. The nominate subspecies, known as the Indian python (*P. m. molurus*), is found from Pakistan throughout India and Sri Lanka to Bengal. It can be distinguished by its lighter color, the 6 or 7 supralabials (upper lip scales) that touch the eye, and the indistinct point to the arrow shape on the head. The Burmese python (*P. m. bivittatus*) is found from Myanmar (Burma) to Vietnam and south to Java, in Borneo and Sulawesi (the Celebes), and in scattered locations along the India-Nepal border. The Burmese python is usually darker than the other subspecies, has a well-demarcated arrow on the head, and has a small row of subocular scales separating the eye from the labial (lip) scales. Although not recognized as a subspecies by most authorities, the Sri Lankan form (*P. m. pimbura*) is darker than the Indian python and has very irregular brown blotches on its back. Some scientists think that the Burmese python should

Burmese pythons (top) have a prominent arrow-shaped mark on the top of their head. Tough, iridescent scales (bottom left) protect the python's body. Burmese pythons have vertical pupils (bottom right).

be recognized as a full species (*P. bivittatus*) because it differs from the other Indian pythons in coloration and scale arrangements. The distribution of the two forms where they come into contact is poorly known. If they do not interbreed, then full species status for Burmese pythons should certainly be considered.

The species and subspecies of the Indian python have a number of common names. The Indian python has also been called black-tailed python, Asian rock python, and Indian rock python. In Hindi, the Indian python is called *ajgar*. The common name "Indian python" is usually reserved for the nominate subspecies (*P. m. molurus*), and *P. m. bivittatus* is known as the "Burmese python." If recognized, the subspecies *P. m. pimbura* is typically known as the Pimbura python or Sri Lanka python. In this section we use the term "Indian python" to represent the species as a whole (*P. molurus*), and "Burmese python" to represent the subspecies *P. m. bivittatus*; when referring strictly to the Indian python subspecies, we follow the common name with its scientific name, *P. m. molurus*.

GEOGRAPHIC RANGE

The geographic range of the Indian python is somewhat poorly known, and it is certainly safe to assume that the species is no longer found throughout much of its historic range because of human activities and development. The locations where subspecies demarcations occur are especially poorly known. Robert Reed and Gordon Rodda (2009) included a detailed description of the geographic distribution of the species in their report on giant constrictors, and we draw from their report extensively here.

In Pakistan, Indian pythons apparently occur only within the watershed of the Indus River and its tributaries. Indian pythons occur throughout Sri Lanka and essentially all of India with the possible exception of the highest elevations in the northern part of the country and the deserts bordering Pakistan. Specimens from various locales all along the Nepal-India border show characteristics of the Burmese python subspecies, and either the Burmese or the Indian python subspecies can be found throughout the lowlands of Nepal in the foothills of the Himalayas, possibly at elevations as high as 8,200 feet. The distribution in Bhutan is poorly known, but the species probably does not occur in the higher elevations. Indian pythons occur throughout Bangladesh and most of Myanmar (Burma), where the subspecies designations are not entirely clear. Burmese pythons are found throughout most of Thailand,

Burmese and Indian pythons occur in temperate areas, including the foothills of the Himalayas in Nepal and northern India.

NATIVE DISTRIBUTION OF INDIAN AND BURMESE PYTHONS

(*Python molurus molurus* and *P. m. bivittatus*).

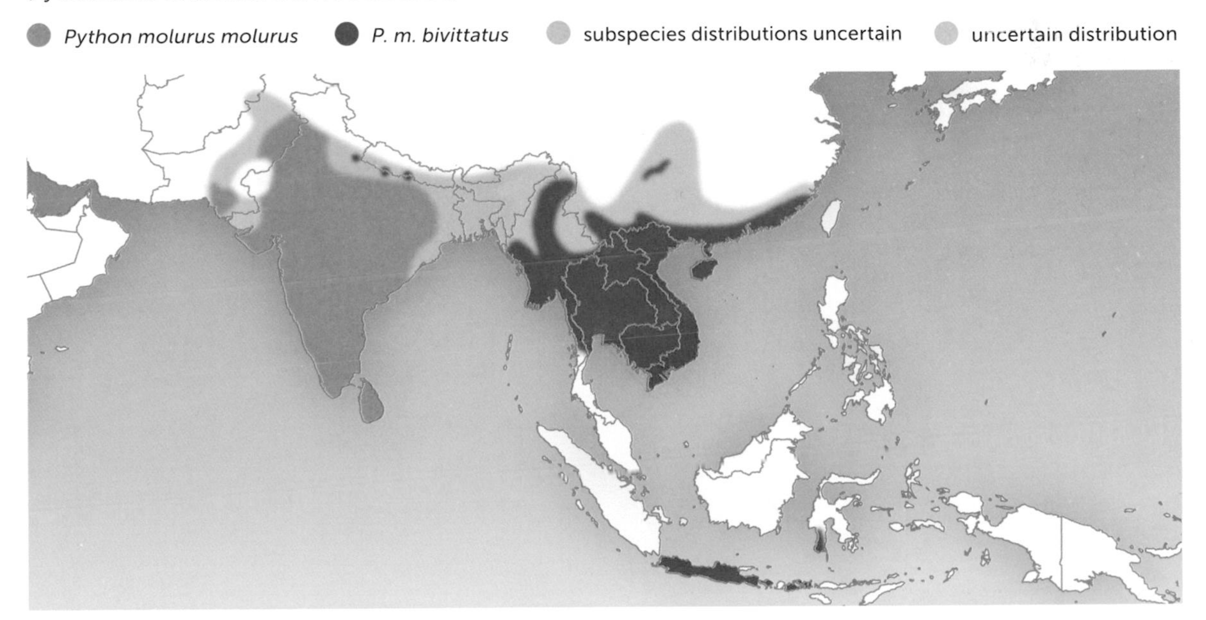

A young Indian boy shows off his pet python to his friends and the camera.

Laos, Vietnam, and Cambodia. The distribution of Burmese pythons in China is not completely known, but the northernmost records of the species in Sichuan province appear to be disjunct (isolated from other populations). Burmese pythons are not found on Borneo, but they do occur on the islands of Java, Sumbawa, and in western Sulawesi.

Some authorities consider the Sri Lankan form of the Indian python to be a distinct subspecies, *Python molurus pimbura*.

EVOLUTIONARY HISTORY

Snakes are a relatively recently evolved group of reptiles. Fossils of snakes are rare, though, and knowledge of the stages of their evolution is thus somewhat lacking. Snakes probably evolved from lizard ancestors approximately 120 million years ago during the Cretaceous period. The evolution of pythons is not well understood either, but pythons are clearly members of a group that diverged from the ancestor of modern snakes relatively early in snake evolutionary history. The fact that pythons, like boas, retain vestigial pelvic girdles and have tiny external hind limbs is a clear indication of their relatively primitive status among snakes.

Evolutionary relationships within the family Pythonidae are not well established. The Indian python (*P. molurus*) is certainly most closely related to the African rock pythons (*P. sebae* and *P. natalensis*), and viable hybrids of Indian and African pythons have been produced in captivity. A single species probably once occurred throughout Africa and southern Asia and was separated by desertification of the Middle East. The Indian python is clearly not as closely related to the other pythons within its range. Some authorities place the large reticulated python and the Timor python (*Python timoriensis*) in a separate genus (*Broghammerus*). Whether herpetologists will accept this designation remains to be determined. Hans Jacobs and his colleagues (2009) reassessed the status of the Burmese python and suggested that it be raised to full species status on the basis of long-recognized morphological characters. Most authorities will probably reserve judgment until new evidence (e.g., molecular studies) is available and the interactions of the two currently recognized subspecies in the areas in which they come into contact are more clearly understood.

African rock pythons are close relatives of Indian and Burmese pythons.

SIZE

The length of snakes length has always been a favorite topic of scientists and laypeople alike. The fear many people experience when they encounter a snake often results in substantial exaggerations of snake sizes. For example, stories of 15-foot (4.5 m) rattlesnakes that stretched completely across a road are common in the southeastern United States, despite the fact that even the largest rattlesnakes rarely (or never) exceed 7 feet (2.1 m) in length. Arguments have gone on for years about which snake reaches the greatest length, and a definitive answer may never be known. Giant snakes are notoriously difficult to measure, and you can always stretch a snake just a little bit more. Measurements taken from shed skins or snake hides are usually inaccurate because skins can stretch considerably.

At hatching, Burmese pythons are small enough to fit in the palm of one's hand.

Most authorities agree that the green anaconda (*Eunectes murinus*) is the heaviest snake in the world. Stories of a 37-foot (11.2 m) anaconda have circulated for years, but most knowledgeable scientists believe it unlikely that anacondas ever exceed 30 feet (9.1 m) in length. The reticulated python (*P. reticulatus*) is probably the longest snake in the world. The exact size reticulated pythons attain is debatable, but in rare cases they may approach 30 feet in length. The fact that a long-standing reward, now worth $50,000, for a live 30-foot-long snake has never been claimed is testament to the fact that snakes of that size, if they occur, are extremely rare.

Although Burmese pythons clearly do not grow as large as anacondas or reticulated pythons, they are considered to be among the largest snakes in the world. At hatching, Burmese pythons are typically 1.5 to 3 feet (45–90 cm) long and weigh less than half a pound. They can grow remarkably quickly if food is readily available, however, and can exceed 10 feet (3 m) within just 2 to 3 years. Large adult female Burmese pythons are typically around 15 feet (4.6 m) long; large males are about 12 feet (3.6 m) long. The largest Burmese pythons certainly exceed 20 feet (6 m) and may approach 25 feet (7.5 m) in extremely rare circumstances. A captive-reared Burmese python was reported in 2003 to measure 27 feet (8.22 m) in total length and to weigh approximately 400 pounds (181 kg). This animal likely represents the longest and heaviest recorded specimen.

The large size of giant pythons and anacondas is an adaptation that not only allows them to feed on animals unavailable to smaller snake

Burmese pythons are among the largest snakes in the world.

species but also likely affords considerable protection from potential predators. Large size may be beneficial for other reasons as well, including those related to thermoregulation (regulating body temperature). Although large snakes may take a long time to warm up, most appear to use physiological and behavioral mechanisms to control their heating and cooling rates much more effectively than smaller species, thus potentially allowing large snakes in temperate climates to remain active and alert when cool weather forces other snakes to retreat underground. For some very interesting reading on sizes of giant snakes, we highly recommend *Tales of Giant Snakes: A Historical Natural History of Anacondas and Pythons*, by J. C. Murphy and R. W. Henderson (1997).

Indian pythons hibernate in porcupine burrows in temperate parts of their range.

HABITATS

Although most people envision pythons living in lush tropical rainforests, both subspecies of the Indian python are found in a wide variety of habitats across their range. Numerous accounts of this species in the literature indicate that it is found in nearly every habitat from lowland tropical rainforest to scrub-desert to temperate forests and grasslands, although no comprehensive habitat study has been conducted of this

Indian pythons inhabit dry regions (top) in northwestern India. Burmese pythons inhabit mangrove forests (bottom left) in many parts of their range. In some parts of their native range Indian pythons take refuge in mammal burrows (bottom right) during the cooler months.

Python habitat in northern India often includes thick vegetation and seasonal ponds.

species across its entire range. Burmese pythons are frequently found in brackish water habitats such as mangroves and marshes, and apparently can tolerate seawater for a time, although they are not typically found in marine habitats. In Pakistan and western India the species is found in very dry habitats characterized by cactus-like plants and few trees. Both subspecies are found in montane areas of the Tibetan plateau and in the foothills of the Himalayas in Nepal, where they may occur at elevations up to 8,200 feet. When water is present, pythons will remain submerged for extended periods. In temperate areas they often use underground retreats such as mammal burrows as refuges and for hibernation. They have been documented in urban areas, but human development has likely limited their populations and resulted in local extirpations of the species throughout much of their historical range. Such urbanization and development in the United States may limit the spread of pythons.

Cris Hagen and Kimberly Andrews observed this Burmese python along the India-Nepal border in Katerniaghat Wildlife Sanctuary.

BEHAVIOR AND PHYSIOLOGY

Few studies have documented anything substantial relating to the behavior or physiology of Indian pythons in the wild. Like most other snakes, however, pythons require relatively little food compared with similar-sized birds and mammals, and many of their behaviors and physical attributes are directly related to this low-energy approach to living. Because they are cold-blooded, or ectothermic, Indian pythons do not require food to fuel metabolic heat production as birds and mammals do. Pythons can literally go for more than a year without food, but they also have the ability to eat frequently when prey is abundant. University of Alabama biologist Stephen Secor and Jared Diamond used Burmese pythons as model organisms to examine the functioning of the digestive system in animals and found that infrequent feeders such as pythons can essentially shut down their digestive system between meals. Immediately after a python swallows a prey item, the snake's metabolic rate increases dramatically and the digestive system prepares to process the food. The intestine can more than triple its weight within a day or two, and many other organs and organ systems increase their activity in order to process the meal. Once the meal is digested and absorbed and extra food is stored in the form of fat, the metabolic rate and digestive system return to resting levels and the intestines and other organs shrink back to their normal size.

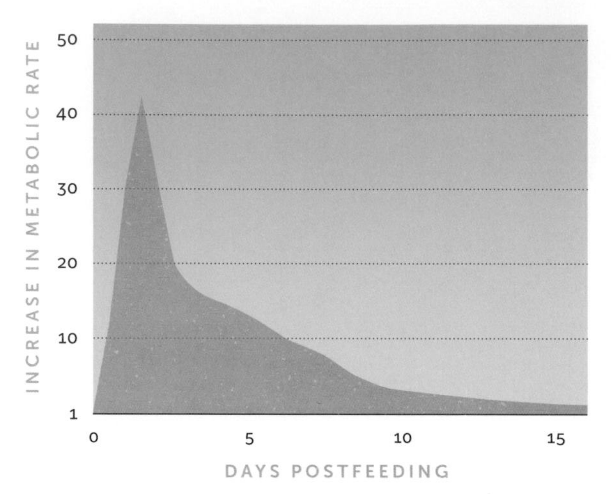

Increase in a python's metabolic rate following consumption of a meal weighing 100 percent of its body mass. Note that metabolic rate increases to more than 40 times that of a resting snake within one day following a meal. Adapted from Secor and Diamond 1998.

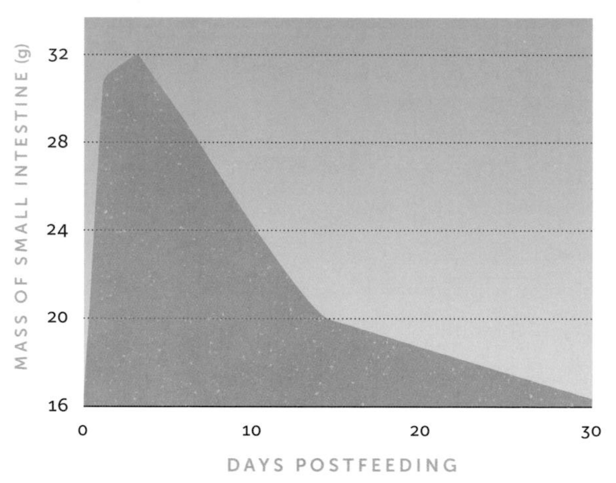

Increase in a python's intestinal mass following consumption of a meal weighing 25 percent of its body mass. Intestinal mass doubled over the course of a single day to allow for digestion and absorption of the large meal. Adapted from Secor and Diamond 1998.

Pythons are generally not very active, at least when compared with birds and mammals. They typically stay in one place for long periods, sometimes months. After consuming a meal, an individual may not move at all for days while the food is being digested. Pythons may move short distances on a daily basis, for example from a burrow to a basking site, but typically, longer-distance movements are rather limited. The extent to which they bask or remain concealed varies with locale and climate. Likewise, the times at which snakes are actively moving about likely varies across their range and according to season. In cooler climates snakes are more likely to bask during the day to increase their body temperature and to be more active at dusk and during the evening before their body temperature drops appreciably. In warmer climates snakes probably limit the amount of time they spend in the sun to avoid overheating but may be active anytime at night if temperatures are suitable. Individuals often remain submerged at the edge of a body of water with only their nostrils above the surface while waiting for an unwary mammal or bird to venture close enough for capture. Such ambush tactics fit well with a low-energy lifestyle. Pythons are also excellent swimmers and often hide underneath submerged vegetation and debris.

In their study of the ecology of Indian pythons in Keoladeo National Park, India, Subramanian Bhupathy and V. S. Vijayan discovered that pythons frequented aquatic areas. They were often associated with porcupine burrows, hollow trees, and termite mounds, typically overwintering within these refuges and spending warmer periods either in water or in thick vegetation. Snakes became relatively inactive during the hottest part of the summer. Indian pythons occasionally climb trees, as well, especially in tropical rainforests, although whether they do so to forage for birds and arboreal mammals or for some other reason is unknown.

Burmese pythons are listed as "near-threatened" in their native range. Pythons may now be more common in Florida than in many parts of the species' range in Asia.

This Indian python is taking refuge in a mammal burrow.

In many areas Burmese pythons could be considered semiaquatic. This Burmese python quickly retreated into water when it was discovered.

Pythons spend much of their time hidden in thick vegetation or underwater.

In higher latitudes and elevations, Indian pythons are thought to hibernate for 3 to 4 months during the winter, often in mammal burrows, and may emerge from their refuges during warm sunny days to bask, as other snakes that live in temperate climates do.

FOOD AND FEEDING

People are particularly fascinated by what snakes eat, in part because they swallow their prey whole and can eat remarkably large meals—sometimes consuming prey that weigh more than themselves. More typically, prey weigh 10 to 30 percent of the snake's weight. The literature includes numerous accounts of Indian python prey. Most records are derived from dissection of snakes, although there are a few accounts of investigators actually finding snakes in the process of killing or consuming prey. Verifying feeding accounts is much like verifying size records: not all of them are accurate and unbiased, and remarkable prey are more likely to be reported than common prey species such as rats and rabbits, even though the latter are probably more important components of pythons' diets.

The vast majority of prey items documented for Indian and Burmese pythons are birds and mammals, including peafowl, pheasants, pigeons, ducks, and other semiaquatic birds as well as rats, flying foxes (large bats), jackals, porcupines, langur mon-

Pythons consume a wide range of prey in their native range. This Indian python (left) is swallowing a chital deer. Pythons may wait in shallow water (below) to ambush prey that passes within striking distance.

keys, wild boar, pangolins, and antelope. Among the deer species reported as prey are barking deer, hog deer, axis deer (chital), chevrotain (mouse deer), and the large sambar deer. Research conducted in Keoladeo National Park in the 1980s documented several bird and mammal species as prey, including spotbill duck, purple moorhen, coucal, cotton teal, grey partridge, redstart, five-striped palm squirrel, rufous-tailed hare, chital, and porcupines. An account of a Burmese python consuming a leopard from the early 1900s is also generally accepted. Domestic animals consumed include goats and dogs. A few reports indicate that pythons occasionally eat reptiles and amphibians, and the literature includes several accounts of Indian pythons consuming large monitor lizards. Fish have been reported as well but are unlikely to constitute a substantial portion of the diet. Indian pythons have been documented to eat carrion, and the guts of several pythons contained fruit. Exactly why a snake would eat fruit is unknown; perhaps the fruit was rotten and the smell confused the hungry snake.

As mentioned above, large pythons are thought to be primarily ambush predators, often remaining in one spot for long periods waiting for prey to approach. A good ambush location is critical for hunting success, and pythons frequently wait along trails made by mammals. They almost certainly use smell or chemoreception to select ambush sites, but likely use vision and thermal cues supplied by the infrared-sensitive pits on their lips to detect approaching prey. A report of snakes in ambush posture moving their tail may indicate efforts to distract or attract potential prey. Pythons may feed on nesting birds, especially in rookeries, and on young mammals in burrows or nests. Such predation would clearly represent active foraging rather than ambush hunting tactics.

Like other snakes, pythons locate prey by using their forked tongue to "taste" the air.

REPRODUCTION

Because the Indian python has a vast geographic range that encompasses many climates and seasonal regimes, making generalizations about reproduction is difficult. Few studies even mention reproduction in this species. In temperate regions the breeding season begins in late winter and may extend through the summer. In Keoladeo National Park in India, pythons mated while congregated around winter refuges. Mating frequency was highest in February and March, and often several male pythons accompanied each receptive female.

Pythons mate in much the same way that other snakes do. The male rubs himself along the female's back, often contracting his muscles, and may use his spurs (vestigial hind limbs) to stroke her. When the female is receptive, she will raise her tail, exposing her cloaca and allowing the male to mate. Parthenogenesis (reproduction without a male) has been documented in captive Burmese pythons, but not in the wild. Regardless, it appears to be an adaptation that allows females to reproduce when males are unavailable.

Reported clutch sizes for Indian and Burmese pythons range from 8 to 107 eggs. Nearly all of the data, however, are from captive snakes, which may be fed unusually large amounts of food and may thus

Indian pythons may breed around winter refugia.

produce larger than normal clutches. Clutch sizes between 20 and 50 eggs are probably more typical of wild snakes. After laying, the female python wraps her body around her eggs to brood them and may raise the temperature of the clutch through shivering thermogenesis. During thermogenesis, the female contracts her body muscles about once every 2 seconds. The warmth generated by the muscle contractions provides an elevated and fairly constant developmental temperature for the eggs within her coils. Frequency of reproduction in the wild is unknown and probably depends on the amount of food available to female pythons. If food is abundant, females may be able to produce a clutch of eggs every year, especially in warmer climates. When food is limited, female pythons probably reproduce only once every 2 or even 3 years.

Female pythons "shiver" to create heat that increases the temperature of their eggs.

PREDATORS AND PARASITES

Documentation of predation on pythons is essentially nonexistent. Eggs and juvenile pythons are certainly vulnerable to predators, but few animals are capable of killing and eating a large adult Indian python. Bengal monitor lizards (*Varanus bengalensis*) prey on python eggs and probably eat hatchlings as well. Crocodilians eat snakes, and where their geographic ranges overlap, probably prey on pythons they encounter in the water or at the water's edge. Large crocodilians can subdue and devour even adult pythons. Birds of prey such as eagles and large hawks can pose a threat to smaller pythons, as can large egrets and herons. Some researchers have also reported pythons apparently trampled and killed by large ungulates.

Several publications report on internal parasites found in dissected snakes, typically round or flat worms (see Further Reading). Whether

In many parts of their range pythons are revered rather than feared. Visitors to this temple in India make offerings to live pythons.

National Geographic's Brady Barr displays a large Indian python captured in its native range.

Mugger crocodiles probably prey on pythons in areas where their ranges overlap.

these parasites are detrimental to snakes in the wild to any substantial degree is unknown. External parasites such as mites and ticks commonly attach themselves to snakes, including pythons. Aquatic leeches may attach themselves to pythons that spend considerable amounts of time submerged.

RISK TO HUMANS

People pose a much greater risk to pythons than pythons do to people. Pythons in their native range are regularly killed out of fear or for their meat or hides. Although large pythons are quite capable of killing adult humans, and even consuming them, they seldom do so. Humans frequently mistake snakes' defensive behavior for aggression, but even large wild pythons are unlikely to bite or attempt to bite humans for any reason other than self-defense. Most accounts of unprovoked attacks on humans involve reticulated pythons or African rock pythons. Nearly all accounts of human deaths attributable to Burmese or Indian pythons involved captive animals. Some occurred as a feeding response when the snake mistook the keeper for prey. A baby reported killed near Hong Kong in the early 1900s was likely predation by a wild Burmese python. A somewhat less believable account tells of a Burmese python stealing and eating the corpse during a funeral. Given the dense populations of humans in Asia and their proximity to python habitat, it is remarkable that pythons have not killed more humans. Of course, it is certainly possible that in remote regions people who go missing are sometimes victims of unreported python attacks.

3 Scientific Research on Pythons

Carefully conducted scientific research on the ecology, physiology, and behavior of invasive Burmese pythons in South Florida is the only method by which we can answer questions relating to their impacts, the risks they pose to as yet unaffected areas, and methods for their control. Sound decisions relating to management of pythons should rely on unbiased, well-designed science. Unfortunately, decisions with wide-ranging effects are sometimes based on outlandish claims perpetuated by the media, politicians, and various agenda-driven individuals and organizations. Basic biological knowledge of the animals can be combined with historical information to understand the conditions surrounding the introduction and establishment of invasive pythons and may be helpful in designing strategies to prevent future introductions of similar species.

In this chapter we describe the methods used to study invasive pythons in the United States and many of the people involved in python research. Most of the information in this chapter and the next is a result of their work.

National Park Service biologist Skip Snow is the point person in Everglades National Park for all things related to the python issue.

WHO'S WHO IN INVASIVE PYTHON RESEARCH

Ray W. (Skip) Snow is directly responsible for much of what we know about python biology in South Florida. He is the primary biologist for the National Park Service coordinating research on invasive pythons in Everglades National Park and its environs. Skip has been the point person for the National Park Service on the python issue since the problem became apparent. He collaborates with numerous scientists and agencies concerned about the invasive python issue and has published several scientific papers and a book chapter describing various aspects of python biology and the risks these snakes pose to native ecosystems. Skip has worked extensively with Mike Rochford (University of Florida), Laura Wilkins (University of Florida Museum of Natural History), and Carla Dove (Smithsonian) to develop a more complete picture of the diet of pythons. Many newspapers and magazines have published articles about him because of the notoriety the python issue has received. Perhaps most notably, *Maxim* magazine selected Skip in 2010 as one of the six people with the "Greatest American Balls" for "catching the 13-foot-long, alligator-swallowing bastards, armed only with a pair of tongs, a laundry sack, and two of the biggest balls this country has ever seen."

Bob Reed of the U.S. Geological Survey has extensive experience working with many snake species.

Ron Rozar of the U.S. Geological Survey works on Key Largo to control python populations.

Robert (Bob) Reed and Gordon Rodda are biologists with the U.S. Geological Survey (USGS) whose research specializes on invasive snakes, and they have published numerous papers on the subject. Both men have worked extensively with the invasive brown treesnake (*Boiga irregularis*) on Guam. In 2009 they published a paper that used climate matching to evaluate potentially suitable climate for Burmese pythons in the United States and analyzed the potential risks several species of large constrictors pose to the United States. Bob Reed is also collaborating with USGS biologist Ronald Rozar to study and remove pythons on Key Largo, Florida. Pythons were first discovered there in 2007 and may pose a significant threat to the endangered Key Largo woodrat.

Frank Mazzotti is an associate professor at the University of Florida's Fort Lauderdale Research and Education Center who is best known for his research on crocodilians in South Florida. His lab is currently involved in several projects related to invasive pythons, most notably a long-term radiotelemetry study examining the spatial ecology of pythons in and around Everglades National Park. In 2010 Frank led a study describing the effects of a severe cold spell on South Florida pythons. Notable members of his lab include Michael Cherkiss, who helps to manage much of the ongoing research; and Michael Rochford, who is responsible for much of the radio-tracking and other research activities being conducted on pythons. Frank Mazzotti's lab, in collaboration with Bob Reed and Kristen Hart of the USGS, has been leading efforts to develop effective trapping methods for pythons.

Kristen Hart is a research ecologist with the USGS who works from the Southeast Ecological Science Center in Davie, Florida. She is well known for her research on sea turtles and diamondback terrapins and is coordinating and leading several collaborative projects on invasive pythons. She currently works with other USGS, federal and state agency, and university scientists to study python mercury levels, stable isotope values, salinity tolerance, genetics, and the risk pythons pose in other parts of the United States. Kristen works regularly with personnel from the National Park Service, the University of Florida, and the authors of this book.

Walter Meshaka, the senior curator at the State Museum of Pennsylvania, has been conducting research on invasive species in Florida for decades. He and his colleagues published an authoritative book on exotic amphibians and reptiles of Florida in 2004 and were the first to conclude that a reproducing Burmese python population is present in Everglades National Park (see Meshaka and colleagues 2000, 2004).

The authors of this book are also extensively involved in python research. Michael E. (Mike) Dorcas, who is a professor of biology at Davidson College in North Carolina, has collaborated on radiotelemetry studies of pythons and is leading studies of their thermal biology. He collaborates with researchers from the USGS, the National Park Service, the University of Florida, the University of Georgia, and Auburn University on various python-related projects. John (J. D.) Willson of the University of Georgia's Savannah River Ecology Laboratory is working on several projects relating to pythons, including construction of models to understand python population dynamics and to evaluate how and when they became established in South Florida. Along with Dorcas, he is also evaluating pythons' abilities to survive in temperate regions and to control their body heating and cooling rates.

Scott Goetz, a biologist with the U.S. Geological Survey, works on Key Largo.

Numerous employees of state agencies, research technicians, and volunteers—too many to be named here—have provided crucial assistance with various aspects of the python research described in this book. Michael Avery (University of Florida) and his colleagues conducted research on the effects of cold weather on pythons; Alex Pyron (City University of New York) and his colleagues used climate-matching studies to examine the potential for range expansion in pythons; Kenneth Krysko (University of Florida Museum of Natural History) has been

Michael Rochford has been extensively involved in many aspects of python research.

Bart Rogers (Auburn University EcoDogs), Melissa Miller (Auburn University), and Lori Oberhofer (Everglades National Park) are shown testing the ability of dogs to locate pythons.

Most pythons that are removed from the Everglades are measured and examined by Skip Snow or other python biologists.

involved in several aspects of python research and curates important biological specimens; and Christina Romagosa (Auburn University) studies python population characteristics and the use of dogs to detect pythons. Other technicians and volunteers who have helped significantly with python research include Paul Andreadis, Chris Gillette, Scott Goetz, Cris Hagen, Bobby Hill, Toren Hill, Josh Holbrook, Trey Kieckhefer, Melissa Miller, Tony Mills, Sean Poppy, and Alex Wolfe.

Scientists tracking pythons by radiotelemetry in the Everglades often need to use helicopters or airplanes.

CAPTURE AND REMOVAL OF PYTHONS

Many pythons have been captured and removed from Everglades National Park and its environs since it became apparent that Burmese pythons were established there. As python encounters increased in the early 2000s, Skip Snow had the foresight to recognize the value of obtaining as much information as possible from each python observed or captured. He instituted a python hotline for people to report information on python encounters and encouraged people to bring pythons into the research lab for examination and to provide detailed information on the exact location and circumstances of the capture. Pythons are typically found by driving along roads and by searching canal banks. All of the pythons captured for Skip's work are eventually euthanized using approved humane techniques. Then they are measured and, in most cases, dissected. During dissection, investigators determine the

Paul Andreadis

If someone says they are "going cruising," it might sound like a high school kid driving the strip to look for eye candy. But not if that person is a herpetologist. We systematically drive roads that pass through suitable habitat, hoping to intercept reptiles and amphibians as they cross. Only by logging lots of miles in the right places at the right times of day can road cruising yield a handsome payoff. I have been road cruising South Florida for the last three years looking for pythons in places where they haven't been documented before.

One night I was doing 45 mph on Tamiami Trail in Collier County, Florida. The Trail is a road cruiser's heaven on asphalt. Running south from Tampa, it cuts eastward at Naples and runs across the peninsula, skirting the northern edge of Everglades National Park on its way to Miami. This section is an especially important venue for python recon. Pythons moving north or west from the Everglades have to cross the Tamiami Trail.

I was cruising a stretch that had not been recently mowed; the grass was maybe a foot high right up to the asphalt. I passed what looked for all the world like a fist and wrist extended out from the grass at the road's edge. "Oh, it *can't* be," I said out loud. There were no headlights in sight, so I slowed and turned around. Sure enough, the "fist" was still there. I passed by at maybe 25 mph, then turned around again. For this third approach I eased up at 10 mph. When I got close, the "fist" clearly turned to the side and disappeared into the grass. I pulled off, grabbed my spotlight, and sprinted after it. The light revealed a large Burmese python crawling down the embankment. When I drew close, the snake bolted, its foreparts rearing off the ground. I waited for it to slow down to a crawl again, then approached. I placed my foot gently but securely on its head, then grabbed its neck. Inspection revealed it to be a female python 8 or 9 feet in length.

Many cruisers are finding roadside pythons in South Florida. What stands out to me about this encounter is that the python clearly was waiting for my car to pass. Given the placement of a python's eyes, it could see in both directions down the highway. That snake had stopped to look both ways before crossing! I do not think that even a major road would, over time, present a barrier to such cautious and patient animals.

Ron Rozar and Mike Rochford (top left) extract a large female python from a root hole in Everglades National Park. Capture and removal of large pythons (bottom left) often requires several people. This python (right) was too large for any available containers.

animal's reproductive condition by measuring the gonads and, if it is a female, the egg follicles. Investigators also examine the entire digestive system for the remains of prey items.

At least 25 species of birds have been recorded in the diet of pythons in Florida, including herons, rails, ibises, and wood stork.

DIET ANALYSIS AND IMPACTS ON NATIVE FAUNA

Invasive pythons affect South Florida ecosystems primarily through predation on native mammals and birds. Concerns center on the pythons' impacts on threatened and endangered species and fears that reductions in populations of common species may affect food webs and ecosystem processes. Nearly all of the information discussed here comes from pythons removed from Everglades National Park and other areas of South Florida. The digestive system of pythons is remarkably powerful. Generally, the snakes digest every bit of the prey except for hair, scales, or feathers, which can remain in a python's gut for extended periods, perhaps even months. Food recovered from the stomach is usually only partially digested and can be identified relatively easily. By the time the food reaches the lower digestive tract only hair or feathers remain, and identification may require specially trained researchers. The next chapter presents detailed information on the diet and impact of pythons in South Florida.

Many pythons captured in South Florida contain the remains of some prey. This python (left) had eaten an opossum. The remains of a great egret were recovered from this python (right).

Dissection of pythons and identification of prey provide considerable insight into pythons' likely impacts, but such studies have inherent biases. First, the prey recovered from a python represents only one or a few of the meals that snake ate during its lifetime. Although it is reasonable to assume that the snake had eaten similar prey at other times, knowing this for sure is impossible. Additionally, investigators are able to sample only a small part of the overall python population, and that sample may or may not be representative of the whole. Most of the pythons encountered and captured in the Everglades are medium-sized subadults and young adults. Whether the prey found within their guts is representative of the diet of snakes in other size classes—larger and smaller—that are less frequently encountered is unknown. Larger pythons probably eat large prey unavailable to medium-sized snakes, and hatchlings probably eat the same prey that many other snake species of the same size eat.

In many cases, a python's stomach contains only tiny bits of its prey's bones, teeth, or feathers.

If we are to fully understand the impact pythons have on native fauna, we must first know about the population densities of prey animals in Florida—how common they were before pythons became established and how common they are now. Unfortunately, adequate data on pre-python densities are not available for most prey species, and most perceived python impacts are based on anecdotal observa-

Radiotelemetry allows biologists to locate pythons hidden in dense vegetation.

tions. Efforts are under way to assemble and analyze historical data on potential prey species that was collected for other purposes. Information on actual python densities would likewise permit sound estimates of the number of prey consumed. That is, if researchers know how frequently pythons consume various prey (based on dissection) and how many pythons are present in the ecosystem, they can estimate how many prey items pythons are consuming. Unfortunately, although scientists studying invasive pythons agree that python populations are large, determining actual densities and how they vary across the landscape is extremely difficult. This issue is discussed in more detail below.

Techniques using stable isotopes may provide additional insight into python diets and impacts. Stable isotopes are forms of basic elements that vary in atomic mass. When predators feed, they incorporate some of the isotopes present in their prey into their own bodies. Generally, predators that feed higher on the food chain (that is, those that eat larger and more complex prey) have higher levels of certain isotopes than those lower on the food chain. Thus, comparing the isotopes present in pythons with those found in potential prey animals can provide clues about the overall diet of pythons in South Florida.

SPATIAL ECOLOGY

Spatial ecology deals with animals' movement patterns and habitat use. Many types of data can be used to address questions related to spatial ecology. For example, the change over time in the locations of python captures within South Florida provides evidence of where introductions may have occurred and the directions in which the population is spreading. Data such as locations where pythons are found are heavily biased, however; that is, pythons are only found in places where people go. Nearly all of the areas where pythons have been seen or collected are along roads or in areas easily accessible to humans. If you did not consider the biases associated with how the data are collected, you might conclude that pythons have an extremely high preference for roads and roadside habitat. Of course, that is not true. Most python records are from roads simply

The vastness and inaccessibility of many areas of the Everglades complicate efforts to track radio-tagged pythons on foot.

because that is where humans are most likely to encounter them.

One way to reduce the biases associated with using incidental captures is to use radiotelemetry to study the spatial ecology of pythons. Radiotelemetry allows researchers to track the movements of individual animals over time and determine their actual movement patterns and habitat use. It involves attaching a radio transmitter to an animal (or implanting one inside an animal) and using a receiver and directional antenna to find the animal and record its location. Radio transmitters typically transmit in the FM range and produce a pulse or beep every few seconds that can be picked up by a receiver. Each animal that is being tracked carries a transmitter that transmits at a different frequency, just as radio stations transmit at different frequencies. Radio transmitters are usually attached to study animals using a collar or some sort of harness, but such attachment methods are virtually impossible for snakes. Thus, transmitters for snakes are nearly always surgically implanted into the body cavity. The surgery is a very minor procedure. Transmitters weigh only a few ounces, and snakes typically heal and recover without complications. After the snake is released, usually a day or two after the surgery, researchers can track it and determine the GPS coordinates of the snake's location, the habitat, and perhaps details of its behavior. Repeatedly locating snakes with radio transmitters over long periods allows accurate generalizations regarding their movement patterns and the habitats they prefer. Geographical information systems (GIS)—software programs that analyze spatial information such as data collected using radiotelemetry—can be used to determine the home range of pythons, when and how far they move, what types of habitat they select, and how their habitat preferences change from season to season.

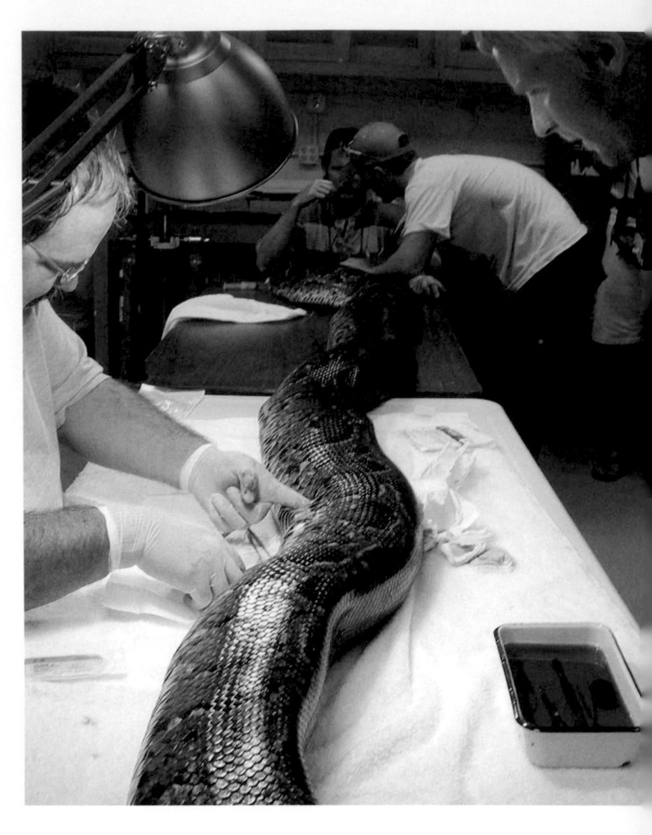

Small, surgically implanted radio transmitters permit biologists to track pythons for months or years.

Members of Frank Mazzotti's lab began a long-term radiotelemetry study of pythons in December 2005 in collaboration with Skip Snow of the National Park Service, biologists from the USGS and the South Florida Water Management District, and Mike Dorcas. It soon became apparent that aircraft (airplanes and helicopters) would be needed to track the animals across the vast, inaccessible habitats they occupy. As of 2010, researchers had used radiotelemetry to study the spatial ecology of more than 25 pythons, gaining considerable insight into many questions relating to the snakes' ecology, behavior, and movement. Researchers are currently exploring the possibility of using satellite-based GPS units to track pythons. Satellite units, which are already used extensively to study sea turtles, whales, and other large animals, send signals periodically to orbiting satellites that relay information to scientists about the exact location of the animal carrying the transmitter. Kristen Hart and Michael Cherkiss have collaborated in studies evaluating the use of satellite-based sys-

A radio transmitter was surgically implanted in this large female python at Davidson College in North Carolina for later study in Everglades National Park.

tems to monitor movements of pythons. The antenna of the satellite transmitter must be exposed, however, and limitations relating to the need to surgically implant such devices in snakes may restrict their use. A 2010 test of a GPS tag without a satellite uplink antenna in a wild python produced promising results.

Radio transmitters must be surgically implanted into the snake's body cavity using relatively minor surgery techniques.

THERMAL ECOLOGY

Temperature affects nearly every aspect of the biology of reptiles, particularly such vital processes as digestion, crawling speed, reproductive cycles, and their ability to sense their environment. Like other ectotherms, pythons must maintain their body temperature within certain limits in order to optimize these important biological functions. They do this primarily through their behavior. The snakes seek refuge from the sun when temperatures are too hot and may bask in the sun when temperatures are cooler. Because pythons are elongated and can be very large, they have a remarkable ability to change the amount of their body surface exposed to the air. If they are stretched out, they expose

Micro-dataloggers about the diameter of a penny are surgically implanted in pythons to give a continuous record of body temperatures.

a very large surface area over which heat can move either into or out of their body. If they coil tightly, they decrease the amount of exposed surface and can retain heat for extended periods. They can probably regulate blood flow to the surface of their skin as well, thus allowing them to retain or more rapidly dissipate core body heat.

Scientists can determine pythons' activity patterns by examining fluctuations in their body temperatures and comparing that variation with environmental temperatures. For example, if a python's temperature is well above air temperature and water temperature, there is a good chance the snake is basking in the sun and absorbing solar radiation. Knowledge of when and under what conditions pythons are most likely to bask provides researchers with information that may help in determining when pythons are most accessible for capture.

Researchers have measured and recorded the body temperatures of wild pythons by surgically implanting dataloggers in conjunction with surgically implanted radio transmitters. The tiny dataloggers are no larger than four stacked dimes and can record and store more than 8,000 body temperature readings. When the datalogger is removed from the python, the data points it contains are downloaded into a computer for analysis.

Data on python temperatures can also indirectly provide insight into the snakes' impacts on prey populations. Scientists can use recorded body temperatures to estimate the metabolic rate of a python at any given time, and from that they can calculate the energy a python requires for its basic metabolic needs over an extended period (e.g., a year). By comparing the amount of energy a python expends with the amount of energy contained in typical prey animals, scientists can determine how many prey items a python must consume to fuel its metabolism.

Climate models can be useful tools for predicting the potential range of introduced invasive pythons. Some climate models predict suitable areas as far north as Washington, D.C.

RISK ASSESSMENT

Researchers have used several approaches to evaluate the possibility that Burmese python populations will expand from their range in South Florida to other regions of the United States. Accurate risk assessments require knowledge of the biology of the species in question, the climates and habitats in which it can occur, and the likelihood that it will be introduced into suitable habitat.

Several researchers have conducted risk assessments based on the likelihood of pythons' establishment and survival. Bob Reed (2005) used ecological and commercial variables to predict the likelihood that various species of boas and pythons would become established in the United States. Ikuko Fujisaki and colleagues (2009) used import records along with assessments of danger to the ecosystem and rate of potential spread to develop a similar risk assessment for a variety of exotic reptiles imported into South Florida.

Climate modeling is another tool that scientists use in risk assessment. Basically, climate modeling compares the climate in the native range of a species with areas outside the native range to determine similarities that indicate regions at high risk of invasion. Climate modeling requires both sophisticated computer programs and scientists well trained in their use. Failure to understand how the variables used in such models affect the results can produce unrealistic and inaccurate projections.

The results of climate modeling used to predict python population characteristics have been controversial. Gordon Rodda, Catherine Jarnevich, and Bob Reed of the USGS published a paper based on climate matching in 2009 that used temperature and precipitation as variables and showed a suitable climate match for Indian pythons (both subspecies) throughout much of the southern United States. The study implied that Burmese pythons now established in South Florida might expand their range to occupy substantially more of the country or that additional introductions of pythons in those areas could result in additional established populations. Numerous people, in particular those with vested interests in the python pet industry (e.g., python breeders and importers), claimed the study was flawed. Alex Pyron, Frank Burbrink, and Timothy Guiher (2008) used a different risk assessment approach known as niche modeling and many more climate variables to conclude that climate suitable for pythons is primarily restricted to the areas of South Florida already inhabited by pythons. Paul Andreadis of Dennison University developed climate models using a different approach that predicted a potential distribution of pythons intermediate between those found by Pyron and his coauthors and Rodda and his colleagues. Results of these models are presented and discussed in the following chapter.

We should keep several things in mind when considering the possible areas in the United States that invasive pythons might occupy. The climate-matching studies that have been conducted used climates from the entire range of the species in Asia.

Scientists kept 10 male pythons from the Everglades in a seminatural enclosure in South Carolina in 2009 to determine whether pythons could survive in a temperate climate.

Mike Dorcas and Whit Gibbons use radiotelemetry to track a python in the South Carolina enclosure.

However, pythons from various parts of their native range probably vary substantially in their tolerance for cold and in behaviors that allow them to cope with varying temperatures. We do not know the origin of the pythons currently established in South Florida. If they are all descendants of individuals imported from tropical locations, they may not have the ability to expand their range into more temperate climates. Additionally, we know little about how cold tolerance might evolve in pythons. Even if the introduced pythons in Florida are tropical in origin and are initially unprepared to tolerate cold climates, future generations may evolve behavioral or physiological mechanisms for cold tolerance similar to those found in animals from temperate regions of the species' native range. The models used to predict suitable climate do not necessarily indicate that pythons will occupy the entire predicted range within a few years or even decades. Even if the models are correct, it may take much longer for pythons to expand their range to the extent predicted. Nor is climate the only thing that may limit the range of pythons. Other factors such as food availability, presence of suitable habitat, or human development may play a role as well, just as they do in the species' native range. We should also consider the possibility that pythons could eventually occupy a greater climatic range in the United States than either climate matching or niche models predict. For example, factors other than climate (e.g., food) may limit their range in Asia. If those factors are not limiting in North America, then pythons could eventually occupy a climate range in the United States greater than that of their native range.

The controversial nature of such modeling and the potential impacts their predictions might have on the establishment of laws and regulations restricting python trade and ownership make it important to determine the accuracy of such models. We collaborated with J. Whitfield Gibbons at the University of Georgia's Savannah River Ecology Laboratory to test the predictions of climate models by monitoring 10 male pythons during the winter of 2009–2010 in a seminatural outdoor enclosure on the Savannah

River Site near Aiken, South Carolina. This location is within the region of suitable climate predicted by Rodda and his colleagues but substantially cooler and more seasonal than South Florida. The enclosure was surrounded by a large overhanging fence and contained numerous brush piles, underground refuges, and a pond. We monitored the behavior, temperature variation, and condition of the snakes several times each week for the duration of the study. Our results are discussed in the following chapter.

STUDIES OF PYTHONS IN THEIR NATIVE RANGE

Our lack of knowledge of pythons in their native range limits our ability to understand their ecology and predict their impacts in the United States. Very few studies have examined the biology of pythons in Asia. Most of the information we have is anecdotal or based on animals kept in captivity. The study conducted at Keoladeo National Park in India discussed in the previous chapter provides some of the best information available on pythons in their native range, but we know essentially nothing about how pythons from temperate climates in Nepal and China differ from those in more tropical locales. Such knowledge would be extremely useful in evaluating the ability of pythons from various sources in Asia to establish populations in the United States and elsewhere. Radio-tracking pythons and monitoring their body temperatures and behaviors in temperate versus tropical localities would be an especially insightful first step in understanding python biology in Asia.

RESEARCH NEEDED

Although we have learned a considerable amount about pythons in South Florida, we still do not understand aspects of their biology that are critical for evaluating the risks they pose to ecosystems. One critical piece of missing information is python population density and how density may vary across the landscape. Determining densities of most wild animals is difficult, but pythons are exceptionally hard to study because of their secretive nature. Typically, ecologists use a method known as mark-recapture to determine population densities. Researchers capture a number of animals within a population, mark them in some way, and then release them. Subsequently, the population is resampled and the number of marked animals captured is compared with the number of unmarked individuals captured to estimate population size or density. Using mark-recapture on pythons in the Everglades is impractical for two reasons. First, it is important that all pythons captured, except for those used in radiotelemetry studies, be removed from the area and not returned in order to lessen impacts on native species. Second, even if we could safely return pythons to the population, the likelihood of recapturing enough pythons in the vast landscapes of South Florida to allow precise estimates of population size would be extremely low. Thus, we need to develop other methods to estimate population densities of pythons.

Another missing piece of information is knowledge of the genetics of the population of pythons currently established in South Florida. Tim Collins and Barbie Freeman of Florida International University and Skip Snow conducted a preliminary study of python genetics in 2008. As of 2010, Kristen Hart was leading a more detailed genetic study that may address some important questions. Are Florida pythons descendants of individuals collected in tropical locations? Are individuals genetically similar (suggesting relatively few initial founders) or quite different (multiple introductions)? Answers to questions such as these are essential if we are to understand the potential for pythons to expand their range and evolve traits that may allow them to withstand cooler climates.

4 Burmese Pythons in the United States

SOUTH FLORIDA ECOSYSTEMS AND THE EVERGLADES

Southern Florida is an environment unlike any other in the world; its reputation as one of America's ecological jewels is well deserved. South Florida is home to a variety of endemic and tropical species that do not occur anywhere else in the United States. The greater Everglades ecosystem, which covers an area approximately 60 miles (87 km) long and 100 miles (161 km) wide across the southern third of the Florida peninsula, is probably the single largest freshwater wetland ecosystem in the eastern United States and is the most extensive region of wet subtropical climate north of Mexico. Bordered on the north by the shallow expanse of Lake Okeechobee, the Everglades is a single vast watershed with shallow water flowing slowly through a "river of grass" extending from the lake all the way to the mangrove forests of Florida Bay. Although human activities have affected the Everglades in numerous ways, much of the region is now protected within a network of parks and preserves that includes Everglades National Park, Big Cypress National Preserve, Collier-Seminole State Park, Fakahatchee Strand Preserve State Park, Biscayne National Park, and Loxahatchee National Wildlife Refuge, as well as numerous smaller public and private conservation lands and property managed by the South Florida Water Management District and the Florida Fish and Wildlife Conservation Commission. Taken as a whole, these lands cover thousands of square miles of largely uninhabited land and represent one of the largest contiguous tracts of protected habitat in the country. The Marjory Stoneman Douglas Wilderness Area within Everglades National Park encompasses nearly 1.3 million acres and is the largest protected wilderness area east of the Rocky Mountains.

Much of the Everglades is a "river of grass" consisting of freshwater marsh vegetated with sawgrass and other emergent vegetation.

The greater Everglades is among the largest areas of mostly uninhabited land east of the Rocky Mountains.

The climate of South Florida is as close to tropical as it gets in the continental United States and is remarkably stable. Average high temperatures range from 75 to 90°F, and average lows range from 55 to 75°F year-round. Although cold snaps typically bring cooler weather

Human-made canals crisscross South Florida. Most were built in the early 1900s to drain the Everglades for agriculture. These canals provide ideal habitat for invasive pythons.

for short periods each winter, freezes are rare. In addition to its warm temperatures, the climate of South Florida is among the wettest in the country, with an average annual precipitation exceeding 60 inches (152 cm), most of which falls during the summer wet season lasting from April to October. The region is periodically inundated by catastrophic hurricanes that sweep in from the Atlantic, Caribbean, or Gulf of Mexico, sometimes leaving massive flooding and destruction in their wake. True to its nickname "river of grass," much of the Everglades is dominated by shallow freshwater marshes vegetated with sawgrass and other wetland grasses. Other habitat types dot the region as well, including *keys* where oolitic limestone outcrops only a few inches high provide enough dry ground for pine trees to take root. Conversely, stands of cypress trees known as *domes* appear to be islands but are actually rooted in water deeper than the surrounding marsh. In southern portions of the Everglades, *hardwood hammock* islands featuring a diverse assemblage of primarily tropical plants provide higher ground than the surrounding sawgrass. Many of these plants, which sport colorful and exotic names such as gumbo-limbo, royal palm, and poisonwood, are found nowhere else in the United States. Toward the southern and western coasts of Florida, freshwater marsh gives way to brackish wetlands that receive saltwater intrusion during dry weather, and finally to vast, impenetrable mangrove forests along the shores of Florida Bay and the Gulf of Mexico. Beyond the mangroves lies Florida Bay, which encompasses miles of shallow flats and mangrove islands stretching to the Florida Keys.

Because of its size, productivity, and warm climate, South Florida is one of the most important areas in the United States for wildlife. The region supports many unique birds such as limpkins, snail kites, short-tailed hawks, and purple gallinules, and is home to many of the country's manatees, American crocodiles, and the last population of mountain lions (known locally as Florida panthers) in the eastern United States. Although many species found in South Florida are also found elsewhere in the United States, the rich marshes of the Everglades are important feeding and breeding habitats for wading birds. The region's warm climate makes it an important wintering ground for millions of birds, from waterfowl to warblers. Even birds that do not breed or winter in Florida use the peninsula as a migration corridor and refueling stop in their long migrations from Central and South America to their breeding grounds throughout the United States and Canada. The waters of the Everglades and Florida Bay provide invaluable feeding and nursery habitats for countless aquatic species, from fish and frogs to aquatic insects, American alligators, and many species of aquatic snakes and turtles. The upland areas of South Florida are home to even more species of reptiles, including the threatened eastern indigo snake—the longest native snake in the country—the eastern diamondback

Florida is the stronghold of the eastern diamondback rattlesnake (left), regarded by many as the king of North American snakes. Banded watersnakes (right top) are perhaps the most abundant snakes in the Everglades. Threatened American crocodiles (right bottom) within the United States are found only in southern Florida.

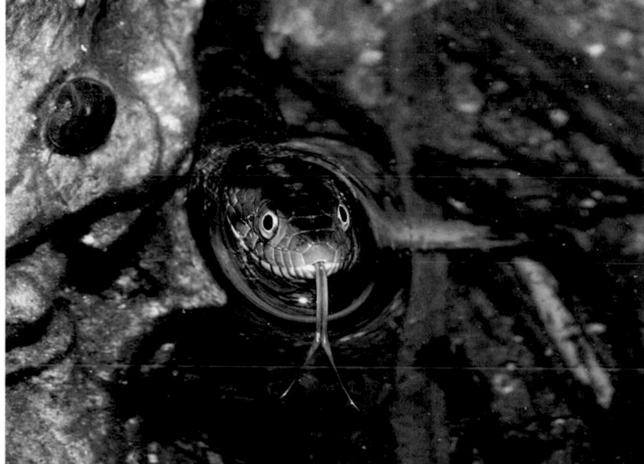

Beautiful veiled chameleons are among the many exotic lizard species established in Florida.

rattlesnake, and the rare rim-rock crowned snake, which grows no larger than an earthworm.

Although the ecological value of South Florida and the Everglades cannot be overstated, human activities have been altering the region for more than a century. Beginning in the late 1800s, hundreds of miles of canals were dug throughout the Everglades, draining the land and allowing conversion of more than 50 percent of the original Everglades to agriculture. South Florida is now an extremely popular tourist and retirement destination, and this popularity has resulted in rampant development, especially along the eastern border of the Everglades on the outskirts of Miami.

The Miami area has also been a hub of the exotic pet trade for several decades, with local dealers receiving regular shipments of a wide array of exotic animals from around the world. The warm climate and presence of exotic pet dealers have made South Florida a hotbed for the introduction of exotic animals. If you walk the streets of suburban Miami, it is not uncommon to see flocks of South American parrots flying overhead, and iguanas and basilisk lizards basking on tree limbs. People fishing in the canals catch fish that are usually seen only in aquariums and pet stores. In fact, of the more than 50 species of lizards found in the eastern United States, roughly two-thirds are established exotic species and nearly all of those are restricted to southern Florida. Of all the exotic species now found in Florida, however, none has gripped the attention of biologists, the media, and the general public more than the rapidly expanding population of Burmese pythons.

The largest and smallest species of snakes in the United States are both introduced exotic species. The Burmese python may be more than 20 feet long, and the Brahminy blind snake usually reaches no more than 6 inches. Both are well established in Florida.

DISTRIBUTION OF PYTHONS IN FLORIDA

Burmese pythons are common pets in the United States, and it is not unusual for escaped or intentionally released pythons to be found anywhere in the country. In warm regions such as Florida, released pythons may survive for long periods without being discovered. In many cases, pythons that originated as captives are recognizable because they are well fed, free from scars or injuries, and are docile when captured or handled. Since pythons have become established in the wild in southern Florida, however, we have learned that some wild pythons also exhibit these characteristics, which makes it difficult to determine with confidence whether pythons found in the wild have been recently released or are part of the expanding introduced population. Despite the certain presence of recently released or escaped pets, however, the distribution of python captures in the region over space and time clearly shows the progression of the established invasive population.

Walter Meshaka compiled early records of pythons in South Florida in his book *The Exotic Amphibians and Reptiles of Florida*, which reports sightings of large pythons in the Everglades as early as the 1980s. Meshaka and others assumed that many of these early records represented escaped or released captives, but the presence of multiple individuals of different size classes by 2000 was sufficient to prompt Meshaka to conclude that Burmese pythons were probably established and reproducing within Everglades National Park at that time.

APPROXIMATE DISTRIBUTION OF PYTHONS IN SOUTH FLORIDA FROM THE 1990S TO 2009

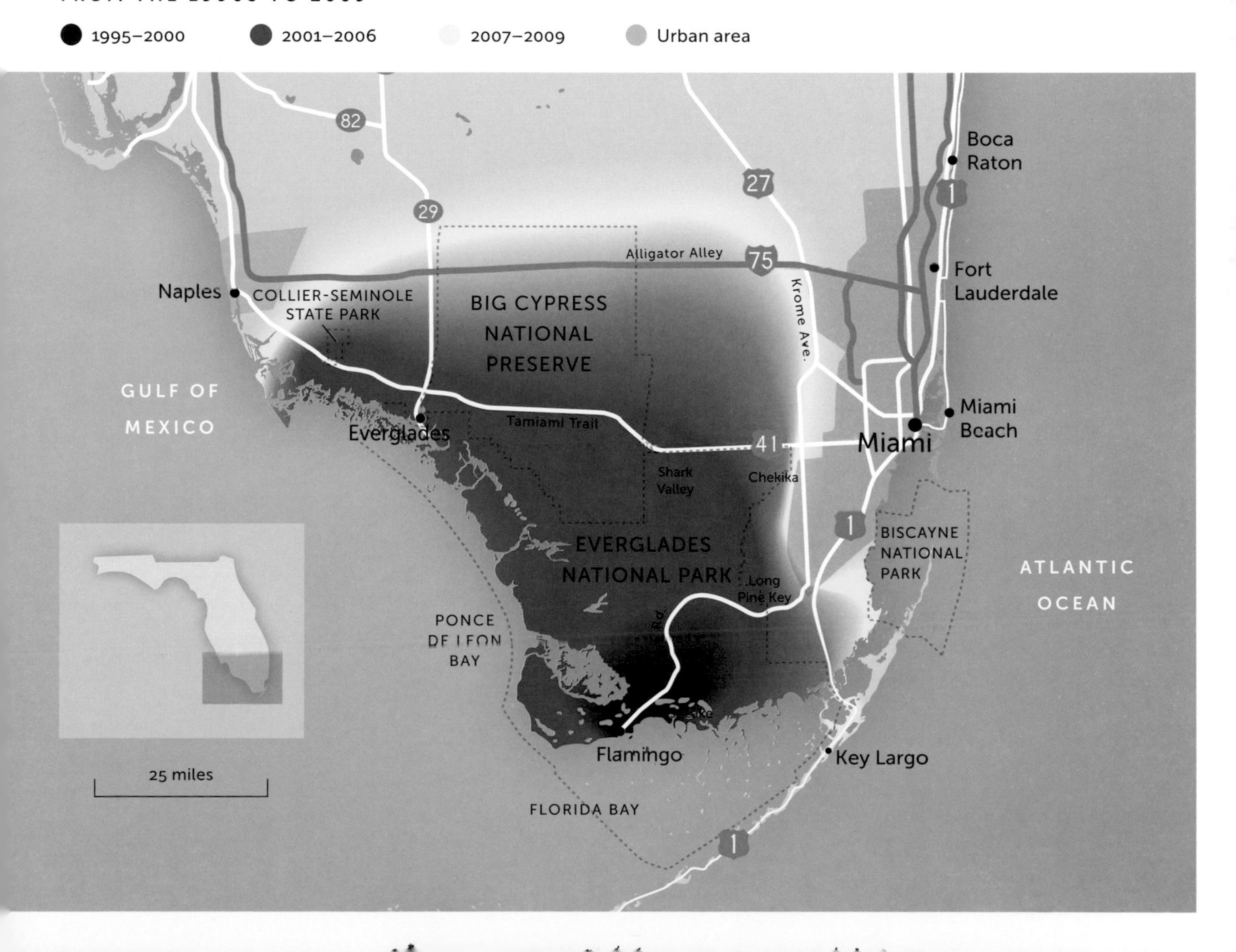

Most of the pythons captured or seen before 2000 were found along the Main Park Road in the mangrove forests just north of Flamingo, although a few were found on Long Pine Key and one individual was found in the Chekika area along the northeastern border of the park. Coordinated searches initiated in the early 2000s dramatically increased the area pythons were known to occupy. Searchers found dozens of individuals, including hatchlings, in the mangrove regions of the southern Everglades, both along the Main Park Road and in remote areas visited only by people fishing. Python records increased in the Long Pine Key area, along the eastern border of the park, at the northern border of the park along the Tamiami Trail (U.S. 41), along the Shark Valley Loop, and along nearby canals. Isolated individuals were found in western portions of the park, in Big Cypress National Preserve, and in Collier-Seminole State Park. A road-killed hatchling that Paul Andreadis found in 2009 within Collier-Seminole State Park provided the first documentation of python reproduction in Collier County. As of 2010, pythons appeared to be ubiquitous in all habitats within Everglades National Park, and numerous individuals had been found in Big Cypress National Preserve, Collier-Seminole State Park, along the western outskirts of Miami, and northward to Alligator Alley (I-75). Individuals have been found in smaller numbers in the vast, inaccessible marshes north of Alligator Alley, along Florida's west coast around Naples, and on Key Largo. Isolated individuals continue to be found well outside these areas, but it is often impossible to tell whether they are released captives or members of newly established or expanding populations. By the time a few individuals are discovered in an area, it is likely that an established reproducing population of these secretive animals is already present. In his report on python reproduction in Collier County, Paul Andreadis suggested that targeted road cruising for hatchling pythons in late summer may be the best way to

Scientists in Florida have examined wild female pythons containing as many as 85 eggs.

Some released pythons may quickly habituate to life in the wild. This albino Burmese python, presumably a released or escaped pet, was found consuming a squirrel in a Fort Myers, Florida, backyard.

Aquatic habitats such as this cypress swamp provide ideal habitat for pythons in Florida.

document areas where pythons have successfully established and are reproducing.

HOW DID PYTHONS BECOME ESTABLISHED IN FLORIDA?

The explosion of Florida python populations has garnered extensive and often sensationalistic media attention. Several scenarios for python introduction have been proposed in the course of this coverage, generally with little or no support or scientific evaluation. They range from paranoid (government scientists released the snakes to prompt bans on keeping snakes as pets) to sensationalistic (hurricanes destroyed giant exotic reptile warehouses and pythons were blown all over the Everglades) to the simple notion that well-meaning but misguided pet owners released their unwanted pet pythons when they could no longer care for them.

Although we will never know exactly how pythons became established in southern Florida, two facts are incontrovertible. First, most or all of the pythons that have been captured in the Everglades since the mid-1990s are wild-born animals, not recently released pets. As of 2010, about 1,600 pythons had been captured across more than 2,000 square miles of southern Florida, some in areas inaccessible to the public. All of these could not possibly be recently escaped or released animals. Second, it would be essentially impossible for Burmese pythons to naturally colonize Florida from their native range in Southeast Asia; thus, the ultimate source of the Florida population is unquestionably the pet trade. Apart from these facts, the circumstances that led to the establishment of pythons in Florida are unknown. By considering the patterns of python captures along with the biology of the species, however, we can evaluate the most plausible scenarios that led to the establishment of invasive pythons in the Everglades.

Although released or escaped Burmese pythons and other large exotic snakes have turned up sporadically in southern Florida for decades, pythons were not encountered regularly in any one area until the mid-1990s. Skip Snow and his colleagues summarized the early history of python captures in the Everglades in a book chapter published in 2007. Between 1995 and 2001, one to three pythons were found each year in the saline glades and mangrove forests of the southern portion of Everglades National Park, particularly between Flamingo and West Lake. The first evidence of wild reproduction by pythons included one juvenile python collected in 1995 and four hatchlings collected in 2002, all near West Lake. As of 2010, pythons were found throughout much of southern Florida but were still most frequently encountered on roads in the mangrove forests of the southern Everglades. This area, which is at least 20 miles (32 km) from the nearest substantial agricultural or residential habitats, appears to be the source of the original introduced population and remains a center of python abundance.

Miami and surrounding areas have been a central location for exotic reptile importers, dealers, and breeders since at least the 1970s, and Burmese pythons have long been a mainstay of the exotic reptile industry. Between 1990 and 2006 alone more than 110,000 wild Burmese pythons were imported from Southeast Asia, and many of them passed through the hands of Florida reptile dealers. Consequently, reptile holding and breeding facilities located north and east of Everglades National Park have been proposed as a primary source of the introduced python population. One particularly popular scenario perpetuated by the media is that pythons were released in 1992 when Hurricane Andrew devastated Homestead and the surrounding area, destroying exotic animal warehouses and releasing their animals. Although the Hurricane Andrew hypothesis sounds appealing and is impossible to completely disprove, we and several other biologists—including Skip Snow, and Kenneth Krysko—question the validity of this idea. We consider it unlikely for the following reasons.

This large reticulated python, which was presumably an escaped or released pet, was found in Miami in 2010.

First and foremost, the area where pythons were first found with regularity—where reproduction was first recorded and the current center of python abundance—is at least 20 miles (32 km) from the outskirts of Homestead and Florida City, where many reptile dealers or breeders were operating before Andrew struck. Pythons did not begin to show up regularly in areas closer to Homestead until well after 2000. Population growth models based on python growth rates, clutch size, and survival suggest that it is not biologically possible for the python population to have grown from a few individuals in 1992 to the thousands of pythons likely present in southern Florida in 2009. Although python breeders could potentially be a source for release of large numbers of young snakes, population growth models suggest that achieving current python population sizes from an introduction in 1992 would require successful introduction of hundreds of young founders. These snakes would have had to make their way more than 20 miles (32 km) from the location of the reptile dealers to the southern portions of Everglades National Park without colonizing intervening habitats (where suitable habitat exists, prey is abundant, and pythons have since become common).

Python population growth models constructed by the authors and published in 2011 and projections based on python capture rates suggest that the most plausible scenarios for establishment of pythons in southern Florida involve a relatively small number of snakes introduced into the mangrove forests of southern Everglades National Park prior to 1985. Whether these snakes escaped or were intentionally released will probably never be known, but that population may have been supplemented by additional unwanted pets that owners released under the misguided assumption that they were "doing the right thing." Pythons likely existed in small numbers for years in the Everglades before people began to regularly encounter them. This is not surprising given the snakes' secretive behavior, the inaccessibility of most areas of Everglades National Park, and the fact that even accessible areas are dominated by thick vegetation and are often seasonally or permanently flooded. As is generally the case with unregulated populations (those with plenty of food and space but few predators or competitors), the python population appears to have increased exponentially between the late 1990s and 2010.

Walter Meshaka documented the presence of Burmese pythons in Everglades National Park in the 1980s.

POPULATION SIZE IN FLORIDA

As of 2010 the number of Burmese pythons inhabiting southern Florida remains unknown. They almost certainly number in the tens of thousands and perhaps even the hundreds of thousands. Skip Snow has

Bob Reed

In early 2008 I spent a morning walking along a levee in the Everglades with my friend and colleague Scott Boback. We carried our fishing gear, intending to alternate between fishing in the adjacent canal and looking for snakes on the banks. Within 15 minutes I spotted a Burmese python partially exposed in weeds next to the canal. I couldn't tell where the head was, so I grabbed a visible coil and heaved the snake farther up the bank. It turned out to be a 6-foot-long male with a bad attitude. We celebrated our luck in finding a python and kept walking, searching, and fishing. Within 20 minutes Scott found another, slightly larger, male python. We were marveling at the odds of finding two pythons in a morning when Scott noticed yet another one a few yards ahead of us. Even with occasional breaks to throw lures for largemouths we ended up catching 6 pythons—all between 6 and 10 feet long—in less than 3 hours.

Scott and I are no strangers to high snake densities—I have worked on superabundant sea kraits and brown treesnakes in the western Pacific, Scott has done research on watersnakes infesting commercial fishponds in Alabama, and we've both done extensive fieldwork on dwarfed boa constrictors that reach high densities on Caribbean islands. While it's possible to catch more individual snakes per day at such field sites than we caught on the levee that morning, I've never had anything like that morning's yield in terms of snake biomass. Searchers on foot catching 100 pounds of snakes in a morning is almost unheard of. I'm in regular contact with biologists at Everglades National Park, and in 2008 I was aware that hundreds of pythons had been removed from the park in previous years, but the density of these big animals hadn't really sunk in. My experience with Scott that day opened my eyes to the ecological implications of our haul, especially in terms of the vast numbers of native animals that the python population must be consuming in the Everglades in order to attain such high biomass in a small area.

By the way, the fishing was lousy.

recorded data from more than 1,600 pythons taken from Everglades National Park and surrounding areas, at least 1,500 of them captured between 2005 and 2010. Despite the large number of snakes removed, pythons continue to become increasingly easy to find in the region. When the authors of this book began studying pythons in the Everglades in 2005, several days of searching might turn up only one or two. As of 2009 we have regularly found multiple pythons crossing roads in Everglades National Park during a single night of road collecting. In fact, we have captured more pythons than rat snakes (a relatively common native species) while road collecting within the park in recent years. Other experienced snake hunters in the region have reported similar trends. Although pythons clearly are now common in many areas of southern Florida and are apparently continuing to increase in abundance, generating accurate estimates of python population size remains one of python researchers' most challenging goals.

Because snakes are so secretive, snake population sizes have been accurately estimated in only a few situations. Clearly, counting every individual python within a large area such as southern Florida is impos-

Even large pythons are incredibly well camouflaged. Here a 9-foot python virtually disappears under a thin layer of pine needles.

sible; in such cases biologists generally try to determine the number of animals within a smaller, well-defined area and then extrapolate this density (number of animals per given unit of area) to larger spatial scales. This approach has been used to estimate numbers of pythons that might be present in Florida based on the few rough estimates of python density determined in their native range. For example, researchers in India counted pythons at winter dens in Keoladeo National Park and calculated an average density of about 13 pythons per square mile, although the authors acknowledged that their method likely underestimated the population size. An equivalent density in Everglades National Park would yield a population of more than 30,000 pythons. A study of African rock pythons, which resemble Burmese pythons in size and habits, found densities in a small preserve in Gambia that were up to 15 times greater than those reported in India (155–194 snakes per square mile). If the Everglades were similarly pro-

PYTHON CAPTURES RECORDED FROM EVERGLADES NATIONAL PARK AND ITS ENVIRONS SINCE 1979

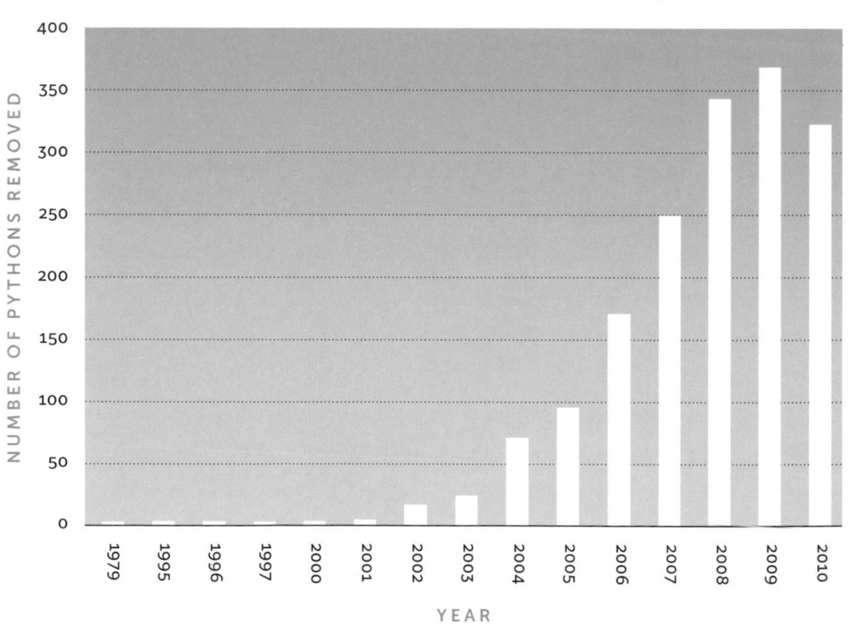

Although it is not uncommon to find pythons crossing roads in the Everglades (left), the number present in the region remains unknown. In many areas road mortality (right) is probably the greatest threat to adult pythons.

ductive, the national park alone could support nearly half a million pythons. Although these numbers are useful as ballpark estimates, they should not be interpreted as the actual number of pythons in the Florida population.

Unfortunately, even extrapolating from small areas of known snake density is not yet possible for the introduced Florida python population. Researchers have found that it is virtually impossible to capture all of the pythons within even small areas, even with very high intensity searches. In fact, although dozens of pythons have been removed from some individual hammocks, canal levees, or road sections within the Everglades, more continue to be found in these areas. Our observations of pythons kept in a seminatural enclosure in South Carolina confirmed how secretive these snakes can be. During warm weather the pythons spent nearly all of their time under debris or underwater and were very difficult to see, even when we could pinpoint the exact location of each snake using radiotelemetry.

A further complication with extrapolating snake density from small areas is that the study area must be representative of the entire area of interest. The area Burmese pythons already occupy in southern Florida encompasses thousands of square miles and many different habitat types. Python density almost certainly varies greatly across habitats, and pythons may be much easier to find in one habitat than another even when densities are similar. For example, nighttime road surveys indicate that pythons are most abundant in the tangled mangrove forests of coastal portions of the Everglades, but pythons are seldom observed in these habitats during the day. Conversely, pythons are commonly observed basking on open, grassy canal banks on cool, sunny days. Which of these habitats actually contains more pythons remains unknown.

When it is not possible to locate all of the animals present within a specified area, biologists generally use mark-recapture studies (described in Chapter 3) to estimate the number of animals present. Using that technique in South Florida would require capturing, marking, and releasing large numbers of pythons back into the wild, however, with no guarantee that they would ever be recaptured—certainly not an appropriate method for a delicate ecosystem like the Everglades where removal of pythons is a top priority. Moreover, even if mark-recapture were attempted in Florida, recapture rates would likely be too low to provide meaningful estimates of python abundance.

Although predators are unlikely to limit python populations, large American alligators are capable of killing most pythons.

In the long run, determining the factors that will eventually limit python populations may be more important than estimating the current python population size in Florida. Such limiting factors might include abundance of predators and territorial interactions among individuals of the same species. Pythons do not defend territories, so territoriality is unlikely to slow python population growth. Likewise, although large American alligators have been seen eating pythons on several occasions, adult pythons have relatively few predators. In most cases, humans are probably the only predators likely to have a significant effect on adult python population numbers. Thus far, removal of pythons by humans, through either directed python management efforts or incidental road mortality, appears to have done little to curb python population growth and spread, at least in the relatively sparsely populated regions of southern Florida. Whether removal by humans is sufficient to keep python populations low in more developed areas such as the outskirts of Miami remains to be determined.

Many dead pythons were found along canal banks following a severe cold snap in January 2010.

Ultimately, prey availability or climate will probably determine the density pythons attain in southern Florida. Although there is some indication that mid-sized mammals have declined in abundance within Everglades National Park since the 1990s, potential prey for pythons, especially wading birds and small American alligators, remains abundant. Could python numbers continue to increase until the snakes have eaten most or all of their prey? This might seem unlikely, but something similar happened on Guam after brown treesnakes were accidentally introduced in the mid-1900s. The population of treesnakes continued to grow until the snakes had driven most of the island's bird and bat species to extinction.

An exceptionally cold period in January 2010 hinted at the potential effects of cold weather on python populations. During this cold event, the most devastating freeze to hit Florida in recent history, Frank Mazzotti and his colleagues documented at least 40 dead pythons in South Florida, including 9 of the 10 large pythons that were being monitored by radiotelemetry. Although this observation suggests that a substantial portion of the Florida python population died during the freeze, Mazzotti and his colleagues also found 59 live pythons during that time. Nobody knows exactly how many pythons died as a result of the cold, and many live and healthy pythons have been discovered since the freeze, including at least 20 hatchlings from the 2010 nesting season. Even if a large percentage of the pythons in South Florida died during the winter of 2009–2010, we know from the rapid rate of the python population increase in the past that the population will probably recover to prefreeze levels within a few years.

Many Burmese pythons in Florida spend much of their time in the water.

A final factor to consider when addressing population limitation is the presence of suitable refuges for snakes. Although the warm climate of the Everglades probably makes underground shelter unnecessary, microhabitats that provide retreats during cold winter weather may be a limiting factor in cooler regions. It is possible that pythons expanding north of their current range in South Florida will survive only in regions where large underground retreats (such as burrows of other species, rock crevices, or caves) exist, and that the abundance of such refuges may dictate python density. In fact, in cool regions of their native range (such as northern India, Nepal, and southern China) pythons are often associated with burrows of animals such as porcupines, which dig extensive burrows that are used by a variety of wildlife.

Kenneth Krysko

For nearly 20 years I have been conducting wildlife research in Everglades National Park (ENP), Big Cypress National Preserve, Southern Glades Management Area, and the Lake Belt region. My graduate studies focused on Florida kingsnakes, and every weekend for two straight years (1993–1995) I walked the canal and levee banks that border ENP (more than 300 total miles). Not once did I see any sign of a Burmese python. Thus, it was a huge surprise to hear that people had been capturing such an enormous nonnative snake species in one of my former study sites. It was difficult to believe, and despite seeing photographs I had to witness it for myself. Our first search during the morning of January 25, 2007, included Florida biologists Kevin Enge, Paul Moler, Skip Snow, Scott Hardin, Brian Camposano, Ellen Donlan, and others, and it took only a few hours to find and remove three live pythons up to 9.2 feet long and two excavated nests of previously hatched eggs.

Several years later I was lucky enough to lead a hunt that set the record with finding the most pythons in a single day and also yielded insight on the susceptibility of introduced Burmese pythons to cold weather. On January 31, 2010, a couple of weeks after a week of record low temperatures, I took a group of friends and some students from my Invasion Ecology of Amphibians and Reptiles class on a daytime field trip to find pythons just north of Key Largo. We split up into four groups, each walking and searching different sections of canal and levee banks for a total of 46 miles in 7 hours, which was also the first large-scale search of the entire area in a single day. We found a total of 14 pythons—7 alive and 7 dead. Pythons were in every section we searched, illustrating that the species is widespread through-

POTENTIAL FOR RANGE EXPANSION

Burmese pythons are well established in southern Florida and their distribution has been rapidly expanding for at least a decade, but just how far pythons will eventually spread and where additional populations could become established remain issues of great concern and considerable controversy. Research described in the previous chapter using climate models offers several possibilities. Gordon Rodda, Catherine Jarnevich, and Robert Reed used climate-matching techniques to predict that much of the southern United States has climate suitable for pythons, as might be expected given that pythons inhabit some temperate regions in their native range. This climate-matching study has been criticized, especially by those in the pet industry, primarily on the grounds that because many captive pythons sicken when kept cool, it simply does not make sense that they could survive outside southern Florida. Subsequently, a paper by Alex Pyron, Frank Burbrink, and Timothy Guiher used a very different modeling approach (niche modeling) and many more climate variables to address possible python range expansion. As is typical of models that include many variables, this study produced a very patchy pattern of suitable climate, with small areas of matching climate scattered

out the area. The smallest snake (approximately 3 feet long) was found dead on top of a levee bank, and the largest snake (more than 13 feet total length) was found alive. Another large python (more than 12 feet long) was found perpendicular to a canal bank with its tail at the water's edge, suggesting that it had used cold shallow water as a refuge but died after it exposed itself to colder ambient air temperatures. Although the cold temperatures apparently killed some pythons (and many other native and nonnative wildlife species), unfortunately it didn't get all of them!

A team led by Kenneth Krysko thoroughly searched an area on the eastern edge of the Everglades following the record freeze in January 2010. The group found seven live and seven dead pythons along 46 miles of canals.

widely across the globe. The only substantial patches predicted in the United States were in extreme southern Florida and Texas. On the other hand, only small patches of suitable climate were predicted throughout Central and South America, which would presumably provide extensive areas of suitable climate for a generalist species like the Burmese python. For example, this study predicted that more than half of Nicaragua would be classified as suitable but virtually none of Costa Rica would be. Additional research by Rodda and colleagues in 2011 demonstrated that removing four blood pythons misidentified as Burmese pythons from the niche-modeling analysis resulted in a climate match encompassing nearly all of Florida, southern Texas, and the lower Coastal Plain of the Southeast. An additional climate-matching study published by Nicola van Wilgen and colleagues (2009) supported the idea that the climate of Florida and much of the Southeast is suitable for pythons and suggested that suitable climate also exists in the coastal Pacific Northwest. All three studies have strengths and limitations, but the true potential for spread of Burmese pythons is likely to remain uncertain for many years.

Problems associated with invasive species cost taxpayers more than $100 billion each year.

Our study of 10 male Burmese pythons from the Everglades maintained in a seminatural enclosure in South Carolina starting in June 2009 (see Chapter 3 and Further Reading) shed considerable light on the ability of pythons to survive in temperate climates. All of the snakes were able to maintain relatively warm body temperatures into early

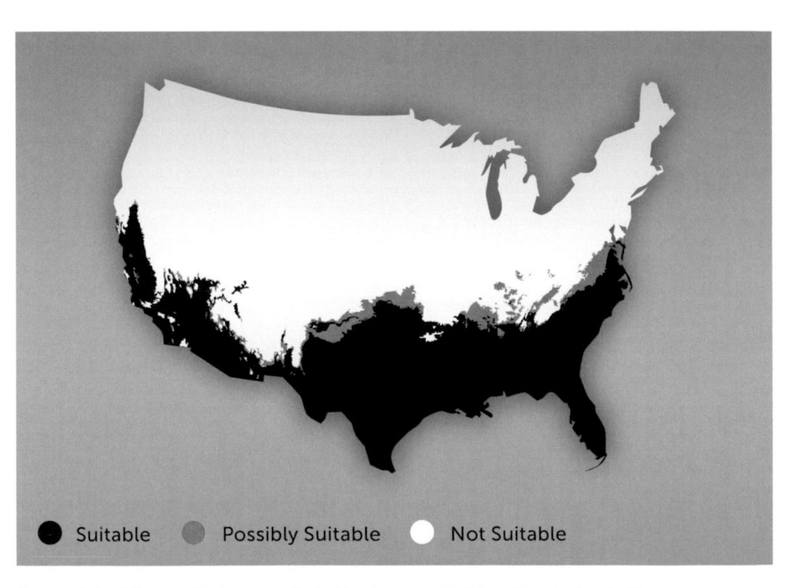

Area of the United States predicted to have suitable climate for Indian or Burmese pythons (*P. molurus*) based on climate-matching models that considered temperature and precipitation. Adapted from Reed and Rodda 2009.

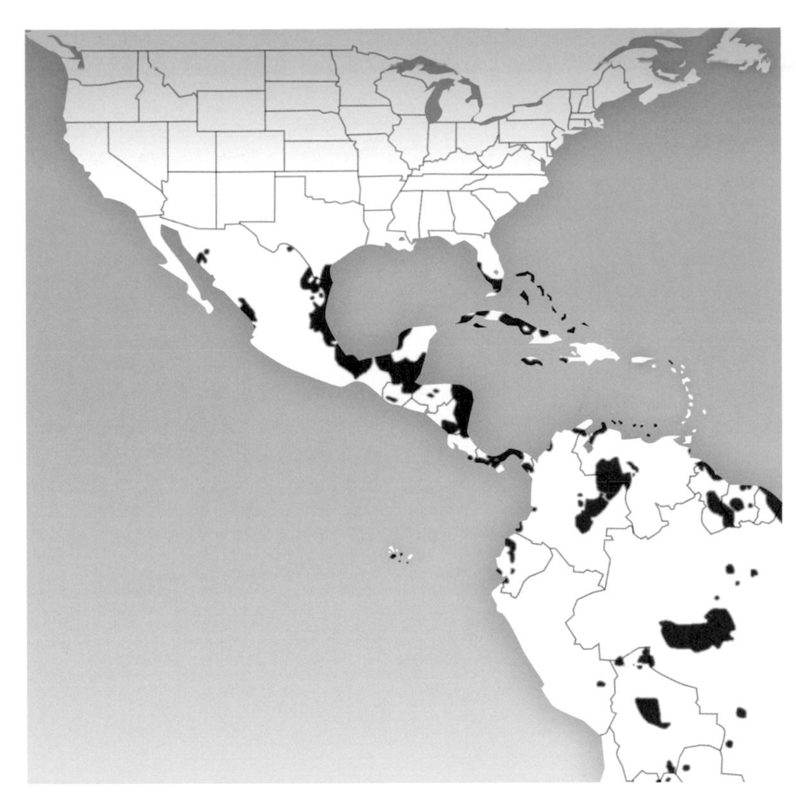

Areas in the Western Hemisphere predicted to have habitat suitable for Indian or Burmese pythons (*P. molurus*) based on niche models that considered 19 environmental variables. Adapted from Pyron, Burbrink, and Guiher 2008.

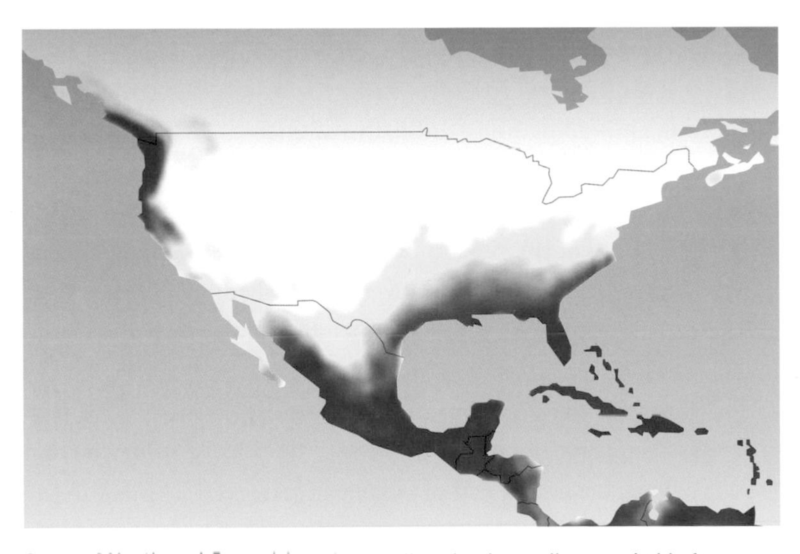

Areas of North and Central America predicted to have climate suitable for Indian and Burmese pythons (*P. molurus*) using a bioclimate approach. Darker colors indicate more suitable climate. Adapted from van Wilgen, Roura-Pascual, and Richardson 2009.

Pythons kept in a seminatural enclosure in South Carolina were able to maintain high body temperatures well into the winter by basking.

A severe freeze in January 2010 killed many pythons in Florida along with native species such as crocodiles, manatees, and sea turtles.

December, and all survived at least 12 nights when air temperatures dropped to between 32 and 40°F (0–5°C), with no apparent ill-effects. At the onset of subfreezing weather in mid-to-late December, however, more than half of the snakes failed to seek appropriate refuges on subfreezing nights and subsequently died. All 10 snakes eventually died, although several survived multiple subfreezing nights and 2 survived until a severe prolonged freeze in January 2010—the same freeze that killed so many wild pythons and native animals in South Florida. In South Carolina this cold snap resulted in a stretch of 14 consecutive nights with air temperatures below freezing and lows of 15°F (-9.5°C) on 2 nights. Both of the remaining snakes sought refuge in artificial burrows during the cold snap but were dead when the burrows were examined after the freeze. Both had unequivocally survived at least 8 nights with subfreezing temperatures, emerging to bask during the day and retreating underground at night. These 2 snakes may have succumbed to temperatures that were simply too cold, or it may be that Florida pythons are not capable of prolonged (2 weeks) hibernation. The results of the South Carolina study suggest that some pythons currently inhabiting South Florida can withstand temperatures appreciably colder than those typical of the Everglades but may not be able to survive severe winters in areas as temperate as central South Carolina.

Frank Mazzotti and his colleagues (2010) reported that many of the

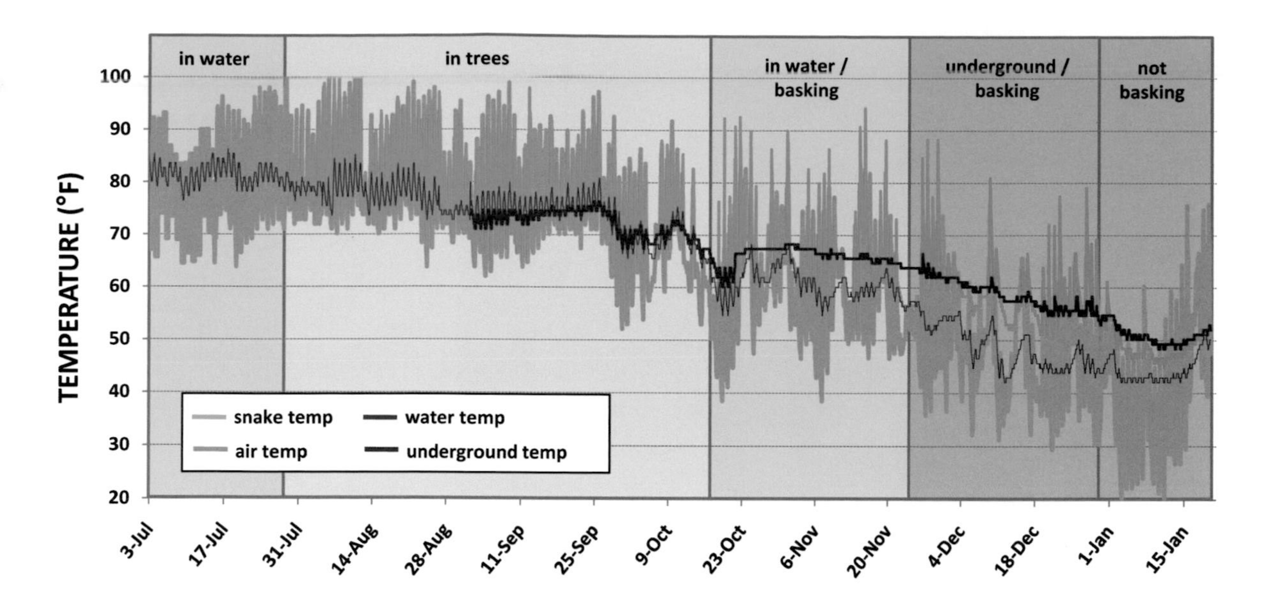

Body temperature of a 7-foot male Burmese python (green) in relation to environmental temperatures (air—gray, water—blue, underground—brown) and the snake's behavior from July 2009 to January 2010 in a seminatural enclosure in South Carolina. Snakes were tracked using radiotelemetry. Adapted from Dorcas, Willson, and Gibbons 2010.

dead snakes they found in Florida after the record cold of January 2010 were coiled on canal banks and apparently had been attempting to bask, even as temperatures approached freezing. Like the South Carolina study, this observation suggests that some of the pythons currently inhabiting South Florida lack the suite of behaviors necessary to avoid subfreezing weather. A substantial number of pythons also survived this cold snap, however, demonstrating that at least some found refuges sufficient to protect them from extreme cold. Michael Avery and his colleagues (2010) reported that pythons being kept in outdoor enclosures in central Florida (Gainesville) also suffered during the freeze; seven of nine died even though they had access to heated shelters. The apparently maladaptive behaviors of pythons in all three areas may reflect the fact that they were acclimated to the climate of South Florida, where freezes are rare. Alternatively, if pythons inhabiting Florida originated from tropical founders, they may not be genetically "programmed" to seek underground shelter during freezing weather. The extreme cold weather in January 2010 may thus have acted as a selective force on the population by killing pythons that did not seek appropriate refuge. Assuming cold-avoidance behavior or physiological cold tolerance has a genetic basis, and thus can be passed from parents to offspring, the surviving population may be more cold tolerant than before.

Finally, although prolonged exposure to cool temperatures often results in terminal respiratory infections in captive pythons, and many people who have kept pythons as pets insist that such infections will limit pythons to South Florida, virtually all of the snakes that died in Florida during the 2010 cold snap and all the snakes in the South Carolina study appeared to have died from acute hypothermia and did not exhibit any symptoms characteristic of cold-induced disease (such as mouth gaping, wheezing, nasal discharge, or congestion in the lungs). Although the 2010 freeze alone was insufficient to eliminate pythons from southern Florida, the observations suggest that cold may limit the extent to which python populations can expand northward; exactly how much remains to be seen.

Pythons kept in South Carolina apparently acclimated well to the seminatural enclosure. This python is feeding on a road-killed opossum.

Although nearly all previous discussions of python range expansion have focused exclusively on cold tolerance, many other factors—including rainfall, habitat characteristics, prey availability, and level of human activity—may contribute to the suitability of other areas of the United States for invasive pythons. Increased knowledge of pythons' cold and salinity tolerances, behavior, genetics, and prey requirements will improve our ability to predict how far this species may spread, but only time will tell just how accurate our predictions are. In the meantime, erring on the side of caution and moving proactively to reduce the chances of python range expansion or additional introductions is certainly the prudent path most beneficial to the environment and human society.

SIZE OF PYTHONS IN FLORIDA

The largest wild python ever found in South Florida was nearly 17 feet long.

Invasive Burmese pythons dwarf the snakes native to the United States and are among the largest snakes in the world. Captive Burmese pythons may top 20 feet (6.1 m) in total length, and some snakes captured in Florida have approached this impressive size. As of 2010, the longest unambiguously wild python captured in Florida was a 16.8-foot (5.12 m) female found by Mike Rochford in Everglades National Park in March 2010 as part of a breeding aggregation that included the large female python and three other males. Several other female pythons longer than 15 feet (4.6 m) have been captured in southern Florida. Although some of the wild pythons captured in Florida have approached the lengths of captive pythons, these snakes generally do not attain the massive weights exhibited by some captives (reported to exceed 400 pounds [181 kg]), which are probably obese as a result of overfeeding and lack of activity. The largest Florida pythons have all weighed less than 200 pounds (91 kg).

Most of the pythons found in Florida since 2000 have been less than 10 feet (3 m) long. The average size of the females, which are generally larger than males, has been at or below the size at which pythons become sexually mature (about 9 feet [2.7 m]). There are at least three explanations for the large number of relatively small, immature pythons we are finding. High mortality of large adults may be one reason, but that seems very unlikely because pythons presumably become less and less susceptible to predators as they get older, and large adults have few predators other than humans. Most of the easily accessible areas in the Everglades are along well-traveled roads, and road mortality or removal by humans in these easily accessible areas might prevent most pythons there from reaching large sizes, resulting in the apparent lack of large snakes. The same cannot be said of the more remote areas of southern Florida, where road mortality is not a factor and very little sampling has been done.

Second, the sizes of captured pythons may not accurately reflect the range of sizes within the overall python population, because snakes of some sizes are more difficult to catch than snakes of other sizes. Biases associated with ease of capture certainly influence our impressions of python size. For example, very small pythons are rarely seen by anyone except snake hunters driving roads at night, probably because young pythons are more secretive and move about less than older ones. Thus, small pythons are probably much more common than these rare sightings suggest. The same may be true to some extent for exceptionally large pythons. Although it would seem that the largest

Burmese pythons are around 2 feet long at hatching, much larger than the young of most native snakes.

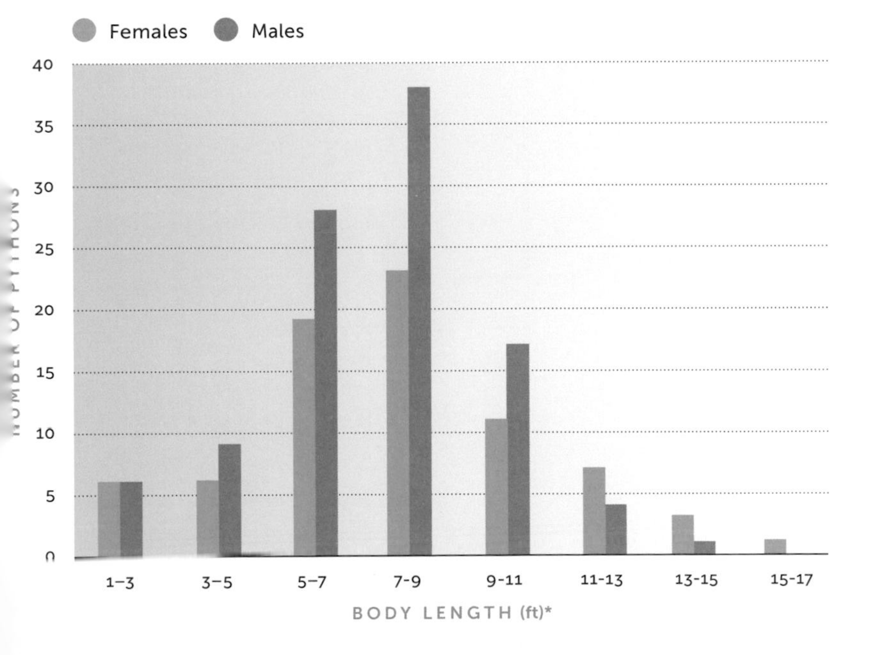

*Males reach maturity at 7 feet, and females reach maturity at 9 feet.

Some pythons are so large that they have been spotted from aircraft.

This python, a 16-foot female captured near Mahogany Hammock in Everglades National Park in 2005, is among the largest individuals captured to date.

snakes would be the most visible, all sizes have excellent camouflage. The largest snakes, however, might move less frequently than smaller individuals or use habitats that are inaccessible to observers (e.g., areas away from roads). Male snakes also generally move more frequently and cover longer distances than mature females, so the smaller male pythons may be encountered more regularly than the larger females.

The third and most probable explanation for the abundance of mid-sized pythons within the Florida population is that the python population in the region is still growing rapidly. A growing population would be expected to contain relatively few large adults and many of their offspring that have not yet reached maturity. Unquestionably, the Florida python population is fairly "young." Because snakes continue to grow throughout their lives, the abundance of large pythons could potentially increase in the future.

DIET OF PYTHONS IN FLORIDA

Determining what pythons eat in Florida is critical to evaluating their impacts on the environment, and diet studies have been a major component of python research since the late 1990s. Mike Rochford and Skip Snow have examined the gut contents of hundreds of pythons from South Florida. With the help of Carla Dove from the Smithsonian Institution and Laura Wilkens from the Florida Museum of Natural History they have identified prey items from many of these specimens and have published several scientific papers and reports on the diets of Burmese pythons in South Florida (see Further Reading). As in their native range, Florida Burmese pythons appear to accept nearly any appropriately sized mammal or bird as prey. They rarely or never prey on amphibians, fish, or reptiles other than American alligators. As might be expected given their large size and rapid growth rate, pythons appear to feed frequently, and most of those examined had some prey remnants within their digestive tract.

Pythons may have already reduced populations of mammals such as raccoons, opossums, and rabbits in Everglades National Park.

Mammals constitute more than half of the prey thus far recovered from pythons, with rodents, especially native cotton rats and introduced Old World rats, being most common. The abundance of rats may reflect the relatively small to medium size of most pythons captured and the fact that many of the largest snakes are captured in the winter when they feed less frequently. Moreover, the reported frequency of rats as prey may be somewhat skewed by the location where the snakes were captured. Skip Snow observed that young pythons from agricultural areas are frequently "packed full of rats"; one snake contained the remains of no less than 14 rats. The abundance of rodents in such habitats may fuel rapid growth of young pythons, and these areas may serve as hotspots of python density, reproduction, and growth. Pythons also frequently consume round-tailed muskrats, rabbits, and squirrels, and have even eaten some endangered Key Largo woodrats. Like all snakes, pythons swallow prey

Large exotic lizards such as this green iguana may provide an abundant food source for pythons in suburban Miami.

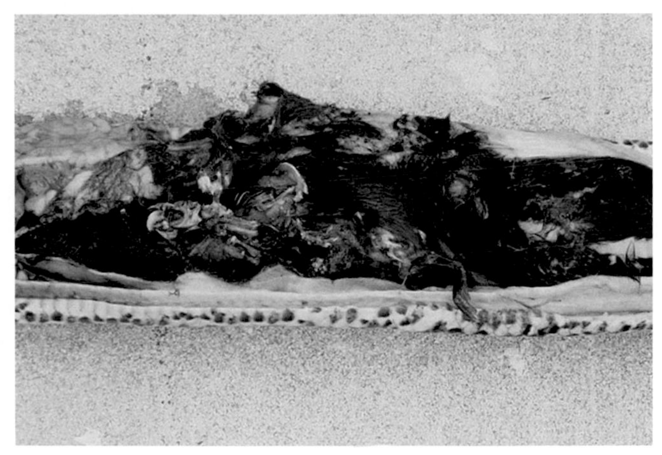

This Burmese python (top), which was found eating a great blue heron in Everglades National Park, is one of the few that have been observed feeding in the wild. Note the heron's sharp bill pressing against the python's neck. The remains of an American coot (bottom left) were found in the stomach of a python from the Everglades. Pythons found in agricultural areas (bottom right) frequently contain the remains of multiple rats in their digestive tract.

Large pythons are capable of capturing and consuming small to mid-sized American alligators.

whole, and larger individuals are able to consume larger prey than smaller snakes. Raccoons, opossums, domestic cats, bobcats, and white-tailed deer have all been documented as prey for large Florida pythons. Birds also constitute a large proportion of python prey in Florida. Marsh birds such as rails, limpkins, pied-billed grebes, ibises, herons, egrets, and wood storks are the species most frequently recovered from digestive tracts. Some of the less common birds recorded are a testament to pythons' opportunistic feeding habits. One python had eaten a house wren, which at an average weight of less than an ounce (28 g) would be a miniscule meal for even the smallest python. Another python contained the remains of a magnificent frigatebird, a species that, although not uncommon in the Everglades, is typically found over the open sea and is generally seen soaring high overhead. How a python managed to catch a frigatebird is unknown, but clearly, hungry pythons will not hesitate to consume any bird they encounter.

Also noteworthy is pythons' ability to kill and eat American alligators. The first account of python predation on alligators involved a dead 13-foot (4 m) python spotted from a helicopter in marsh habitat well away from accessible areas of Everglades National Park. The snake was partially decomposed, missing its head, and had the rear portion of a 6-foot (1.8 m) alligator protruding from its midsection. Dramatic photographs of this sight prompted extensive media attention. Although determining exactly what caused the death of the python is difficult,

This bobcat claw was removed from the stomach of a Burmese python from South Florida.

Pythons often wait in the water to ambush prey, such as this opossum, which they drag into the water before consuming.

the snake clearly had consumed (and probably killed) the alligator. Mike Rochford and Skip Snow subsequently documented several other instances of Florida pythons preying on alligators. Although native lizards are likely too small to be of interest to most pythons, large introduced exotic lizards such as iguanas, monitors, and tegus, which are established in some areas of southern Florida, may become important diet items for pythons in those habitats.

BEHAVIOR AND THERMAL BIOLOGY

Observations of the day-to-day behavior of free-ranging Burmese pythons in Florida are rare. Most of what we know stems from people's observations of where, when, and how they encountered pythons; behavior of wild pythons monitored using radiotelemetry; and observations of pythons that were maintained in a seminatural enclosure in South Carolina. The vast majority of python sightings in the Everglades have involved snakes crossing roads or apparently basking along the banks of the many canals that crisscross southern Florida. Pythons are most frequently observed crossing roads at night during warm weather, suggesting that they are nocturnal for much of the year. Conversely, snakes

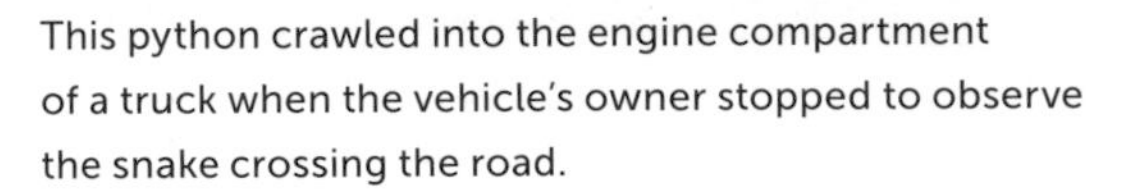

This python crawled into the engine compartment of a truck when the vehicle's owner stopped to observe the snake crossing the road.

This young python was found coiled on a pulley in a water control structure.

are most often observed along canal banks during the winter months, often exposed to the sun and apparently basking to raise their body temperature. In fact, during cool winter weather it is not uncommon for searchers to encounter several basking pythons on canal banks within a few hours. Chance encounters with pythons have occurred in a wide variety of other situations as well. People fishing and airboat operators periodically observe pythons, and several snakes have been spotted in trees. Pythons are also frequently killed or discovered during land-clearing operations such as mowing or disking and are sometimes observed after prescribed burns or wildfires. Several pythons have been found in unusual places, such as inside buildings and water control structures. A visitor to Everglades National Park who stopped his truck to observe a python crawling across the road reported that it attempted to escape by crawling under the vehicle and quickly found its way into the truck's engine compartment.

For several years biologists led by Frank Mazzotti, Skip Snow, Kristen Hart, and Rebecca Harvey have been using radiotelemetry to monitor wild adult pythons in the Everglades (see Further Reading). One initial question that their research was designed to investigate was how much time pythons spend along roads and canal levees. Because pythons were initially captured most often on roads or along canals, it was thought that they might be drawn to such disturbed habitat types. Radiotelemetry studies have determined, however, that pythons actually spend most of their time away from roads and canals. Radio-tracked pythons ranged widely and were capable of moving several kilometers at a time, but they frequently spent long periods in one location apparently inactive or "resting." They used a wide array of habitats, including mangrove forest, open freshwater marsh, forested habitats, and modified habitats such as canals and fallow agricultural fields.

One unexpected by-product of the python telem-

Several male pythons kept in an outdoor enclosure in South Carolina (above) spent a lot of time in trees. Skip Snow and Lori Oberhofer (left) unload pythons captured after a morning spent searching canal banks for basking snakes.

etry studies was an understanding of their homing ability and potential for long-distance movement. When radiotelemetry of pythons in the Everglades was initiated in 2005, several of the snakes that were being tracked were released far from their original capture locations to make tracking easier and to avoid areas frequented by visitors. Within a few months three snakes had moved many miles, following nearly straight paths back to where they were originally captured. One snake moved 48 miles (77 km) back to its capture location within just 2 months, a movement rate of nearly a mile per day (see Harvey and colleagues 2008). Although long-distance movements are not characteristic of normal python behavior, pythons are obviously capable of moving great distances and have the ability to navigate accurately across the landscape. How they navigate has not yet been determined, but homing has been documented in several other snake species as well.

Although telemetry has yielded insight into python movements, much of the tracking has been conducted using aircraft, which seldom allows the observer to actually see the snake or observe its behavior. Even when snakes are located in an area that a researcher can reach on foot, pythons are typically in thick vegetation or in the water and are difficult to approach, let alone see. One fact that has become clear through telemetry studies is that pythons spend the great majority of their time well hidden. They are extremely at home in the water, often remaining in shallow water for days or even weeks at a time.

Pythons can conserve heat by positioning their body in a tight coil. A tightly coiled python can stay appreciably warmer than the air temperature for extended periods.

In terrestrial habitats python often seek shelter underground.

Pythons are capable of moving long distances through the Everglades.

Although pythons in Florida certainly seem to be semiaquatic, determining just how much they rely on aquatic habitats generally is difficult because few locations in the Everglades are more than a few yards away from some sort of wetland.

Temperature dataloggers have been implanted in many of the pythons that have been tracked in the Everglades to monitor their body temperatures throughout the year. Although pythons are "cold-blooded," the dataloggers have shown that they are certainly not cold. In fact, the snakes used behavior to maintain their body temperatures within a fairly narrow range much of the time, generally between 70 and 90°F (21–32°C), with the optimal temperature probably around 85°F (28°C). For much of the year in Florida the snakes were able to maintain these temperatures by hiding under cover during the heat of the day and restricting their activity to the morning and evening. In many cases water provided an ideal thermal environment for pythons, because many aquatic habitats in South Florida remain between 70 and 90°F (21–32°C) year-round, and the water provides a buffer against both hot daytime and cool nighttime temperatures. Even in the cooler months pythons were able to maintain body temperatures within their optimal range most of the time, generally by basking in the sun during cool weather and retreating to water or cover at night. Most of the pythons monitored in Florida between 2005 and 2009 occasionally experienced body temperatures between 40 and 50°F (4–10°C) and apparently suffered no ill effects.

Observations of the 10 pythons that we maintained in South Carolina, which also had implanted dataloggers, generally confirmed the patterns of activity and behavior observed in wild snakes in Florida.

Pythons swim well and can apparently tolerate brackish water.

In the summer the pythons spent most of their time submerged and moved primarily at night. When not in the water, they were coiled deep within brush piles or thick vegetation on land. They were almost never visible, except perhaps for a nose barely protruding above the water or a small section of coil barely visible deep within a pile of brush. To our surprise, some of the snakes spent considerable amounts of time in trees during the summer. Arboreal behavior was most pronounced among snakes less than 8 feet (2.4 m) long, and many individuals spent days at a time coiled high in the branches of pine trees within the enclosure. As the weather cooled in the fall, the South Carolina snakes generally became much more visible. When temperatures dropped below 60°F (16°C), the snakes commonly basked in plain view. By basking they were able to reach body temperatures up to 90°F (32°C) even on days when high temperatures were in low 50s and dropped below freezing at night. At night they generally retreated to the water or deep within brush piles, and most individuals used artificial burrows at night when temperatures dropped near or below freezing.

Feeding and foraging behavior is a particularly poorly understood aspect of python biology. Because pythons feed infrequently in comparison with warm-blooded predators and stay hidden most of the time, opportunities to observe them foraging in the wild are limited. In fact, apart from a few rare instances when pythons have been discovered in the act of constricting prey, virtually nothing is known about how pythons in Florida find their prey. Because pythons are highly aquatic, we assume that they either ambush prey from thick vegetation or shallow water, or slowly prowl through these habitats searching for prey. Observations of pythons in their native range and those maintained in South Carolina suggest that pythons frequently use ambush foraging strategies. The snakes kept in South Carolina spent large amounts of time hidden in shallow water. Although snakes frequently accepted prekilled prey while in these locations, they seldom struck violently and often refused to accept prey unless left undisturbed for long periods. The time of day and seasonal aspects of foraging by Florida pythons are also largely unknown. The South Carolina snakes accepted prey during both day and night in warm summer weather but appeared to be most active at night. Presumably pythons constrict and kill most prey before consuming it; the South Carolina snakes frequently dragged prey animals into shallow water before swallowing them.

Although pythons are generally secretive in the wild and docile when raised in captivity, they can be quite irascible when they feel threatened. When approached, even the largest wild pythons usually try to flee, and they can escape to cover or water with remarkable speed. When captured or cornered, however, they readily display a range of defensive behaviors typical of snakes, including loud hissing, emission of copious quantities of vile-smelling musk and feces, and violent striking. Their length allows large pythons to strike several feet in any direction, and their long, recurved teeth are capable of inflicting serious lacerations. When pressed, pythons tire quickly. Chris Gillette, who has captured numerous pythons in the Everglades, reported at least two individuals that apparently feigned death following capture.

REPRODUCTION IN FLORIDA

Skip Snow and his colleagues have gathered a substantial amount of information on reproduction of invasive Burmese pythons in Florida by examining

Chris Gillette

I was walking in some tall grass in Big Cypress when I stepped on something solid in the mushy marsh substrate. I rolled my foot over it tentatively and could feel that it was cylindrical like a log, yet softer. I bent over into the sawgrass and began to part it several inches with my hands until I saw the glimmer of scales below the vegetation. Large scales. An infusion of adrenaline hit me. I wrapped my hands around the snake's body, and at about that moment a head erupted out of the grass 4 feet away and struck at me. I dragged the writhing snake from the grass and, realizing it was a good-sized individual, tossed it into the trunk of my car as it tried to strike back out at me. When I opened my trunk later, a distinctive hiss emitted from inside, followed by a strike that extended outside the vehicle. I grabbed the snake behind the head and pulled it out to measure and photograph. As I was measuring the snake, it was fighting to escape, as snakes usually do; then it stopped moving or fighting and went limp. I let go of the animal after measuring it out right at 10 feet long, and it didn't move at all. Fearful I had somehow killed it, I tapped it on the head with my shoe; no response from the snake. I even tried my bare hand against its head to elicit a response. Nothing. A few minutes ago this snake had been striking and hissing, and now it wouldn't budge—no response at all. I began to talk to my friend about what could be the problem while standing relatively motionless a few yards away from the animal, when one of us noticed the python *very* slowly creeping away from us! I took a few steps toward it and it froze again. I touched it on the head and got no response. So then I tried rolling it over and it

wouldn't let me! The snake automatically righted itself every time I tried to flip it over. I could pick it up and drape it limp over my shoulders. A few hours later we took the snake out again and it was striking, hissing, and acting normal. Once we grabbed it by the head, though, it struggled a few moments then went limp and played dead again.

captured pythons and radio-tracking free-ranging ones. Python reproduction in Florida appears to be similar to that of pythons in subtropical and temperate regions of Asia. Most mating in Florida apparently occurs as temperatures warm in the spring, particularly in March and April. Beginning in late winter, it is not uncommon to find females attended by multiple males in "mating balls," and males probably range widely during this time in search of receptive mates. Females lay their eggs in May and June, and clutch sizes in Florida have ranged from 18 to 85 eggs, with an average of about 40. Nests are well hidden and only a handful have been found, usually by tracking gravid (pregnant) females by radiotelemetry to their nest site. Most of those nests were under some type of cover such as under the roots of trees, in limestone crevices, or under decaying vegetation or discarded trash. As in their native range, females coil around their eggs and actively defend them if disturbed. In the summer of 2006 Skip Snow used a remote camera to observe a nesting female python that Bobby Hill, an experienced python collector, had found in a burrow. Skip was not only able to film nesting behavior but actually documented the female convulsively contracting her muscles to generate heat to warm the eggs. This behavior, known as shivering thermogenesis, had never before been filmed in the wild. This ability to warm the eggs may allow Burmese pythons to nest successfully in temperate climates. Thus far, only one of the python nests biologists have located has been allowed to hatch in the wild—and only after a fence was built around it to keep the hatchlings from escaping. The nest contained 22 eggs, 17 of which hatched

This "mating ball" consists of several males and a single female.

Python nests discovered by biologists are generally removed.

AIR TEMPERATURE AND NEST TEMPERATURE DURING THE PERIOD A FEMALE PYTHON WAS BROODING HER EGGS

Note the relatively warm nest temperature during a relatively cool period from July 13 to July 16; the female probably kept the eggs warm using shivering thermogenesis. Adapted from Snow et al. 2010.

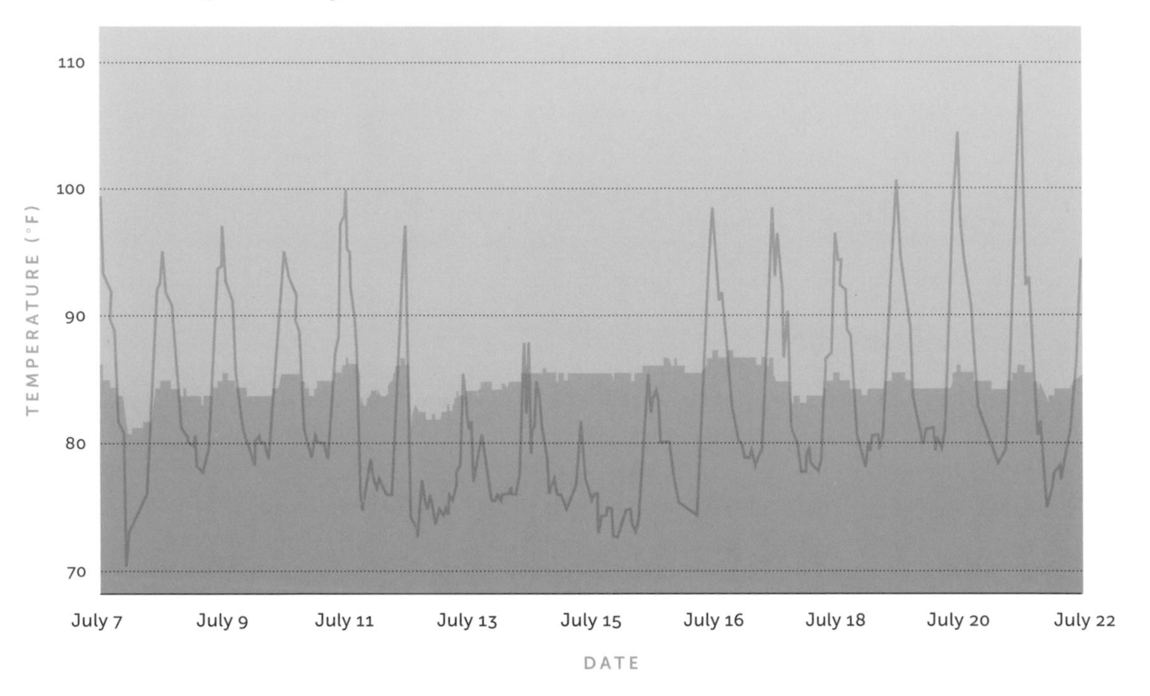

This female python is coiled around her eggs.

This female python captured in Florida contained many eggs.

in late July. This observation and the seasonality of hatchling python captures during road surveys suggest that python eggs typically hatch in July and August. The 2-foot (61 cm) hatchlings leave the nest immediately and disperse into the surrounding habitat. Other than a few individuals captured crossing roads, often very late at night, hatchlings are seldom seen. They apparently grow rapidly, and most pythons are about 5 feet (1.5 m) long by the end of their first year. Females reach maturity at lengths of about 9 feet (2.7 m) and probably reproduce for the first time in their third or fourth year of life.

Pythons in Florida have been recorded eating birds ranging in size from tiny wrens to large herons and egrets.

IMPACTS ON NATIVE WILDLIFE

Introduced Burmese pythons represent an entirely novel top predator in the United States. Not only are they much larger than any native snake species, they also grow rapidly, have many young, can occur in high densities, and are capable of eating a broad range of prey. In fact, a large python is capable of consuming nearly any species of mammal or bird that it encounters in the southern United States. Moreover, like many snakes, pythons can survive for long periods without food, presumably allowing them to persist even when prey populations in an area are depleted. Finally, because snakes are so adept at remaining concealed,

Biologists allowed one python nest to hatch in the wild after surrounding the area with a fence and documented the successful hatching of 17 baby pythons in late July.

At least 32 pythons totaling more than 800 pounds in weight were removed from a 5-mile stretch of this canal in one winter.

The brown treesnake was accidently introduced on Guam in the mid-1900s and has since driven many of the island's native birds, bats, and lizards to extinction.

it is possible for large pythons to live unnoticed even in close proximity to people. For example, many large pythons have been captured along a canal that borders one of the busiest roads in South Florida. In one winter alone at least 32 pythons with a total weight of 826 pounds (375 kg) were removed from a 5-mile (8 km) stretch of this canal bank, and pythons are still frequently found in the area. It is a testament to the secretiveness of these snakes that so many have been able to survive and grow large in an area frequented by fishermen, bordered by a busy highway, and regularly patrolled by python hunters.

The ability of introduced snakes to affect ecosystems has already been demonstrated. When brown treesnakes were accidentally introduced onto Guam in the 1950s, they rapidly drove to extinction nearly every bird and bat species native to the island. Despite efforts to control or suppress them, treesnakes remain abundant on Guam, and virtually the only other vertebrates now present on the island are small lizards and introduced rats. Although Florida's birds and mammals are probably

Sadly, this is the closest most people ever get to seeing a Florida panther. Already at risk because of habitat loss and road mortality, endangered panthers may have to compete with introduced pythons for food and could even become python prey themselves.

better prepared to deal with snake predators than those of Guam, which had no large native predators, the case of the brown treesnake illustrates the destructive potential of introduced snakes.

Burmese pythons exist in high densities in southern Florida and are certainly consuming large numbers of native mammals and birds, but their impact on populations of native wildlife has yet to be determined. Stephen Secor and his colleagues presented research in 2006 that determined that a python consumes approximately 3 pounds of prey for every pound of weight it gains. Thus, a 100-pound (45 kg) female python has likely consumed at least 300 pounds (136 kg) of native wildlife over the course of her lifetime. Diet analyses and observations of python behavior indicate that mid-sized mammals and marsh-dwelling birds such as rails, wading birds, and waterfowl are the animals most likely to be affected by pythons. Several scientists, including the authors, have noticed that raccoons, opossums, marsh rabbits, and other mammals that were once abundant in the Everglades are becoming increasingly difficult to find. Josh Holbrook compared the abundance of mammals along the main road in Everglades National Park with similar habitats north of the current range of introduced pythons (see Further Reading). He observed only 1 opossum and no raccoons within Everglades National Park while driving roads at night, and 12 opossums and 21 raccoons in similar habitats where pythons do not yet occur. Bob Reed and Gordon Rodda (2009) suggested that Everglades populations of round-tailed muskrats may also have declined in recent years. Thus, evidence that fewer mid-sized mammals are now found in areas of the Everglades where pythons have been established for many years appears to be mounting. Scientists and managers fear that similar declines may occur in regions more recently colonized by pythons.

The most immediate concern related to python impacts is their effect on rare or endangered species. Both wood storks and Key Largo woodrats are already rare enough to warrant federal protection, and both have been recorded in the diets of pythons. In fact, although pythons have only recently colonized Key Largo, several of the estimated 200 remaining Key Largo woodrats have been found within the stomachs of Burmese pythons. Pythons also threaten many of the marsh and mangrove species that are characteristic of the Everglades, including roseate spoonbills, anhingas, purple gallinules, white-crowned pigeons, "great white" herons, and American alligators. Several other rare species are also at risk from pythons, although their geographic distribution, body size, or habits make them less likely to be python prey. These include the Cape Sable seaside sparrow, key deer, American crocodile, and whooping crane. Although some might think a top predator such as the Florida panther would be immune to predation, large pythons have consumed leopards in their native range, and a young panther is certainly small enough to become prey for an adult python. Thus, we cannot discount potential impacts on this iconic South Florida species.

Pythons might compete with the federally threatened eastern indigo snake, but indigo snakes actually eat a much broader range of prey than pythons and might even eat young pythons.

Although the risks pythons pose to rare or endangered species have received more attention, long-term ecosystem-level changes that could result from the python proliferation are perhaps more alarming. Many species may be indirectly affected by shifts in mammal and bird abundances in areas where pythons become common. For example, Florida panthers, bobcats, and birds of prey could suffer if the small mammals or birds on which they rely for food decline in numbers. It is also worth mentioning that some species may actually benefit from the presence of pythons in the Everglades. For example, reduction in numbers of raccoons and opossums, which often prey on bird and turtle eggs, might result in improved nesting success for turtles or songbirds in the region. The pythons might also eat destructive exotic species such as feral cats, hogs, and nutria. Although some people fear that pythons will compete with native snakes such as the federally threatened eastern indigo snake, the impacts of pythons on native reptiles will probably be minor. Competition does not generally appear to be a limiting factor among snake populations, and by the time pythons are a year or so old they are generally consuming prey animals larger than those regularly eaten by

Bobcats (top) are among the predators that may be negatively affected by pythons, which both eat them and may also compete with them for rabbits and other prey. Once a common sight, it is now rare to see opossums or raccoons scavenging roadkill (bottom left) along the Main Park Road in Everglades National Park. Biologists suspect that python predation may have already caused declines in some Everglades mammals. Marsh birds such as this limpkin (bottom middle) are common prey items recovered from pythons in South Florida. Perhaps the most serious risk posed by invasive pythons is the impact they may have on Florida's rare or endangered species, such as this Key Largo woodrat (bottom right).

Skip Snow

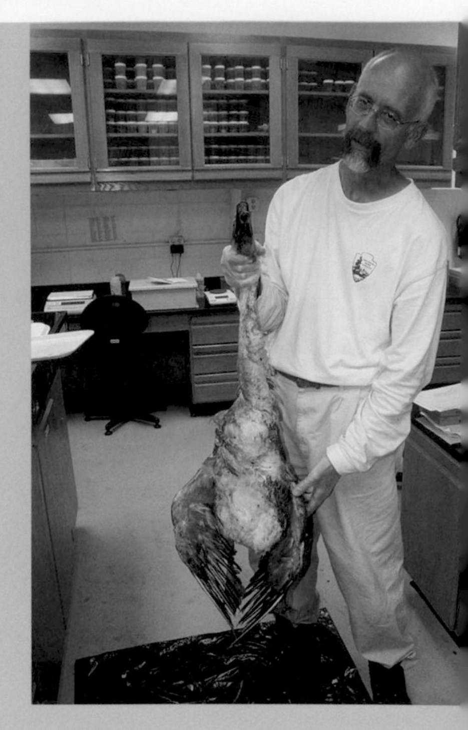

In 2006, soon after we began radiotracking pythons in the Everglades, I documented what was probably the first case of a wild python in Florida incurring a measurable economic cost. Over the course of several weeks, one python that I had been tracking on the edge of national park land made its way into the vicinity of some rural homes. As the snake moved closer and closer to the homes, it became clear that it could become a nuisance and I decided that the time had come to remove the snake from the area. When I tracked the snake on the day it was to be removed, though, I found it had entered the property of one of the homes whose owner kept a small flock of domestic geese in the yard. When I approached the landowner, intending to ask permission to enter the property and remove the snake, he immediately remarked that one of his geese had gone missing. Sure enough, I tracked the snake and found it concealed in vegetation at the edge of the property with an obvious bulge in its stomach, which proved to be the missing goose. I apologized and removed the snake, but the landowner insisted that the National Park Service replace his goose. Not wanting to create an incident, I ordered several goslings as replacements, but when they arrived the owner complained that goslings were not a fair replacement for the adult goose. Luckily, a colleague of mine, Tony Pernas, kept geese and was willing to trade one of his adult geese for the goslings that I was now stuck with. I made the trade and loaded the adult goose into the back of my station wagon, happy just to have the whole situation over and done with. To add insult to injury, however, when I arrived to deliver the replacement goose, I opened the trunk to find that it had defecated all over the rear of my car.

In Asia, Burmese pythons will eat domestic animals such as dogs and goats.

native snakes. A possible exception may be the eastern diamondback rattlesnake, a large snake that preys frequently on rabbits. Although the potential exists for introduced pythons to spread disease to native reptiles, pythons are only distantly related to any snakes native to the eastern United States, and thus far nearly all the pythons examined have appeared healthy with few parasites.

RISK TO DOMESTIC ANIMALS

While the impact pythons will have on native ecosystems remains to be determined, the fact that they can and will consume domestic animals is unquestionable. In their native range Burmese pythons commonly consume pets and domestic animals, particularly poultry and small livestock such as goats. Pythons in Florida have already been recorded consuming domestic chickens, geese, turkeys, and cats. Additionally, the authors have heard several unsubstantiated reports of pet dogs (including a poodle) being eaten by pythons in suburban Miami.

A python Skip Snow was tracking on the edge of Everglades National Park in 2006 approached a home and consumed a domestic goose. The relatively minor inconvenience and cost involved in replacing the goose is probably fairly typical of the types of interactions residents of rural and suburban South Florida should anticipate as pythons continue to spread and increase in abundance in the region. Although very large female pythons could probably consume young livestock, pet dogs, cats, and poultry are probably most at risk. In general, the danger posed to domestic animals will probably be similar to the risk already posed by American alligators in the region, with one important difference. Because pythons are so secretive, they can persist in close proximity to people without being noticed and are not necessarily tied to aquatic habitats. It is quite possible that pythons taking up residence in vacant lots or along the banks of canals will enter suburban yards and consume cats, dogs, and other animals. Such incidents will likely become a high-profile problem as pythons increase in suburban areas of South Florida.

Cats and small dogs are probably the domestic animals in most danger from invasive pythons.

RISK TO HUMANS

Media coverage of invasive pythons has often focused on the possibility that pythons will make predatory attacks on humans. Although the risk of an introduced python attacking and potentially killing a person has probably been blown out of proportion, the risk cannot be completely discounted. Without question a large adult Burmese python has the ability to kill a person and could certainly consume a child. They have been known to do so in Asia, although such events are extremely rare and Burmese pythons apparently attack people far less frequently than their larger cousin the reticulated python.

Perhaps the most telling testament to the power of large constrictors is the many injuries and deaths that have been attributed to captive pythons in the United States in recent years. The Humane Society of the United States has reported at least nine people killed by pet pythons in the country since 1999. Some of the attacks occurred when owners were feeding large pythons, which quickly learn to associate the presence of the owner or the opening of the cage with food. Typically, the snake accidentally strikes the owner, thinking that food is being offered. Unless another person is present to help remove the snake, the

Python meat "does not taste like chicken to me," says biologist Skip Snow. He found it to be "bland, rather chewy, white meat. Not offensive, but generally tasteless and in need of a tenderizing marinade, or slow cooked as in a stew."

python is often able to subdue the owner with powerful constriction, sometimes resulting in death before help can arrive. These types of cases suggest that most attacks by pythons involve the python confusing the owner with prey, a situation that is usually exacerbated by the presence of food that the owner is planning to feed to the snake or the scent of prey on the owner's hands. Several other cases, however, involved pythons killing infants or toddlers who presumably were not attempting to feed them. In 2009 an 8-foot (2.4 m) pet albino Burmese python escaped from its cage in a Florida home and subsequently constricted and killed the owner's 2-year-old daughter. Tragic incidents like these strongly suggest that pythons will at least occasionally view humans as prey items.

Although no deaths had been attributed to wild pythons in Florida as of 2010, Skip Snow knows of at least three instances of wild pythons making apparently predatory strikes at people. In one of these cases several biologists were tracking a large female python through thick sawgrass marsh within Everglades National Park. The snake was so well concealed in shallow water that the group walked past it before

Most wild pythons bite readily when captured or cornered.

realizing where it was hidden. As they passed, the snake struck violently from its ambush position, latched onto the lower leg of one of the biologists, and held on. Fortunately, the biologist was able to pry the snake's jaws apart and, with the help of others, put a safe distance between himself and the snake. He was shaken and sustained bite wounds to his lower leg but was otherwise unharmed. This type of strike is quite different from a typical defensive strike of a Burmese python, which is generally accompanied by hissing and body thrashing, and then quick release. The other two apparently predatory strikes did not result in bites but also involved snakes that were unmolested and unprovoked and people who were unaware of the snake's presence before it struck.

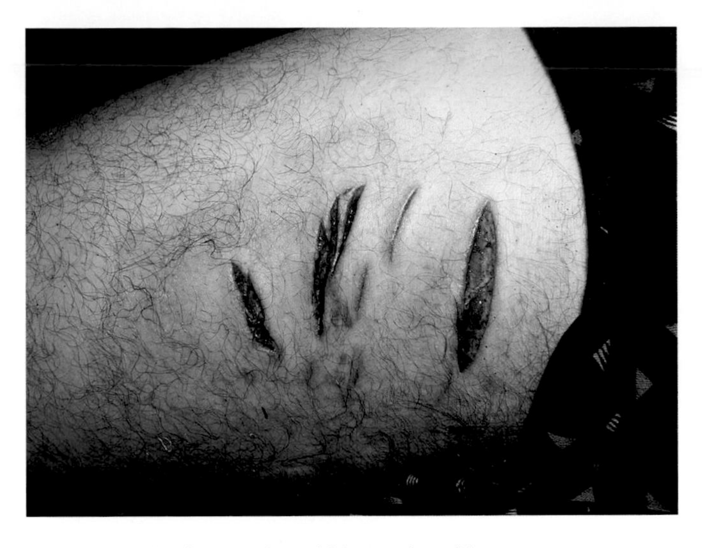

Even a defensive (nonpredatory) bite, such as this bite inflicted by a large wild reticulated python, can be a serious injury.

People who have worked extensively with wild pythons in Florida generally believe that pythons do not normally consider adult humans to be potential prey. Most of the hundreds of pythons that have been captured in the Everglades exhibited nothing but fearful or defensive behavior toward the humans involved. Also, scores of fishermen frequent canal banks where pythons are commonly found, and we know of no substantiated reports of pythons attacking people in these situations. Like the American alligators found in similar habitats throughout the Southeast, most pythons probably pose little threat to people.

Pythons can harm people in a variety of other ways, directly or indirectly, in addition to the obvious case of predatory attacks. Python hunting has become a popular recreational activity in some parts of South Florida, and the state has established an open season for hunting pythons in some areas. Nearly all wild pythons exhibit a relatively nasty disposition when captured, and most do not hesitate to bite defensively. Pythons possess long, sharp teeth, and a bite from a large individual is a serious injury, sometimes requiring stitches. Several recreational python hunters have been bitten and injured by captured pythons as a result of casual handling. Certainly injuries could result from accidents during python hunts, especially when firearms are involved. Additionally, although python meat is reported by some to be quite tasty, recent studies have documented dangerously high mercury levels in the tissues of some pythons captured in the Everglades. Consuming contaminated python meat could result in mercury poisoning or other health problems. Finally there is the potential for python-vehicle collisions or for motorists to crash while trying to avoid a python that is crossing a road. A large python represents a substantial obstacle to a small vehicle. Most pythons cross roads at night in areas close to canals or other water bodies, and it seems likely that vehicle accidents related to pythons may be the greatest hazard pythons pose to Florida residents.

5 Control Methods for Burmese Pythons

The most cost-effective control method for nearly any invasive species is prevention of the initial introduction. Once an invasive species has become established, control is often difficult and costly. Billions of dollars are spent in the United States each year combating invasive plants and animals. Most invasive species are extremely difficult to eradicate; in some cases eradication is essentially impossible. Controlling invasive snakes is exceptionally difficult for several reasons. Although snakes occur in high densities in some areas, they are usually extremely secretive and thus very difficult to detect and remove. The efforts to eradicate invasive brown treesnakes on the small island of Guam is a good example of this phenomenon. Although efforts to control and suppress populations of brown treesnakes have been somewhat successful in certain areas, the fact that they are unlikely ever to be eradicated from an island about one-tenth the size of Everglades National Park demonstrates the difficulty researchers face in eradicating pythons from the vast and largely inaccessible Everglades. It is possible, however, to capture large numbers of invasive snakes in some situations. When the population is reduced in this way, the number of snakes captured over time declines accordingly, but detecting or eradicating the last remaining individuals is often nearly impossible. Likewise, incipient populations are extremely difficult to locate and completely eliminate because at this stage the snakes are present at such low densities that they are seldom encountered.

African rock pythons and boa constrictors are established in small areas of South Florida.

If introductions are detected early, eradication or at least control may be possible with appropriate intensive and coordinated efforts. In 2009, for example, it became apparent that African rock pythons (*P. sebae*) are established in a small area just outside Everglades National Park. Because the population is currently confined to a reasonably small area it may still be possible to eradicate them, but doing so will require quick mobilization, a variety of coordinated approaches, and extremely intensive efforts. Such efforts require substantial funding, and serious efforts to coordinate and fund an eradication program for this species are probably insufficient as we write this book. If the African rock python population continues to expand, eradication or even significant control will likely become impossible.

Federal, state, and local participants have explored or discussed numerous approaches and methods to control the invasive Burmese pythons present in South Florida and to prevent introductions of other species. As the magnitude of the python problem became clear, researchers quickly realized that plans to eradicate the snakes completely from South Florida were unrealistic. Consequently, control efforts are now directed toward developing methods for removing snakes from particularly sensitive areas or at least suppressing their numbers and thus their potential impact on native fauna and ecosystems. In nearly all cases the secretive nature of pythons is the primary difficulty in trying to control them. Bob Reed and Gordon Rodda (2009) have analyzed essentially all of the possible control techniques. Here we draw on their report to discuss some of the methods that have been used, tested, or at least proposed to increase our ability to locate, remove, and possibly even control pythons in critical areas of South Florida.

Pythons typically use rectilinear locomotion, inching their belly scales along and pushing their body forward in a straight line.

SEARCHING FOR PYTHONS

Pythons appear to congregate in certain areas (such as along canals), and manual searches of these areas under the right conditions can sometimes result in the capture and removal of numerous snakes. The percentage of the population actually removed in such efforts is unknown, but even after several years of such removals, numerous pythons are still captured in these areas. Richard Bauer of Davidson College demonstrated pythons' ability to avoid detection while helping to monitor 10 male pythons kept in an outdoor enclosure during the summer of 2009. The enclosure was approximately 100 feet (30.5 m) on each side and contained a pond and numerous shrubs and brush piles. He allowed volunteers to search the enclosure for 20 minutes. Despite the fact that many of the searchers were experienced snake hunters, few were able to locate even a single snake in the enclosure because most of the time the snakes remained either underwater or deep inside brush piles.

Road cruising remains among the most effective ways to find and capture pythons in Florida.

"Road cruising" is perhaps the most efficient way to find pythons. Searchers drive at moderate speeds along roads, often at night, keeping watch for snakes either crossing or at the edge of the road. Snakes are not necessarily attracted to roads—in fact, they may try to avoid them—but because a driver can cover so much ground quickly and snakes are very visible when on the road, the technique is very effective. Road-

Pythons are extremely secretive and well camouflaged. Sometimes the only sign that a python has passed by is a wide track in the mud.

Even after a python is located (top), capturing it can prove difficult. In remote areas of the Everglades, finding a python may be only half the battle. Here (bottom), python biologists Skip Snow and Ron Rozar haul out a large python.

cruising researchers and volunteers are responsible for finding most of the pythons that have been removed from Everglades National Park. There are few roads in many areas of South Florida, however, especially in interior areas of Everglades National Park, and most of the park is essentially unavailable to road cruising. Even if roads crisscrossed the entire Everglades, intensive collecting using road cruising would likely have limited impact on python populations. Road cruisers have been removing pythons along the Main Park Road from the visitor center to Flamingo for several years, with multiple vehicles searching on many nights, but they do not seem to have reduced the python population. In fact, from 2000 to 2009 the frequency of pythons found crossing the Main Park Road steadily increased despite removal of hundreds of pythons.

JUDAS SNAKES

When members of Frank Mazzotti's laboratory and collaborators began to track and study the spatial ecology of pythons in 2005, they soon discovered that tracking a snake often led them to other snakes, especially during the late winter and early spring breeding season. A large reproductive female python may attract numerous males, and in several instances researchers following a radio-tagged snake captured

Jeffrey Fobb

Hunting snakes requires a mixture of knowledge, patience, persistence, and luck. I have been fortunate enough to locate as many as five snakes in an evening (usually a very late evening), but on many nights the time spent is fruitless. It is a struggle against millions of years of evolution resulting in animals well suited to hiding in the extensive habitats of South Florida.

As an example of the serendipitous nature of finding pythons, I traveled to one of the areas that have a well-established population of Burmese pythons for a late-morning search. I was accompanied by an environmental reporter from an international newspaper who wished to document the activities of someone licensed to remove pythons. Unfortunately, we searched for approximately 3 miles but failed to find a python.

I returned the reporter to her hotel and almost immediately received a call from my wife, Sandy. She had remained at home to train one of her donkeys and had noticed that he was balking as he passed a certain area outside our home. When she followed the donkey's gaze she saw a large python in the grass, basking in the sun. With the help of our daughter, Kiera, Sandy attempted to encourage the snake to enter a large dog kennel. She was hanging on to a young donkey that had no interest in approaching the snake and she lacked the proper tools to deal with a surly 9-foot python, but she was successful in safely containing the animal until I returned home with the reporter in tow and was able to capture the snake. The Burmese python, a male approximately 9 feet long, appeared to have been under our home for some time. It may have been attracted by the presence of prey because my daughter has a small flock of chickens. As is generally the case in snake hunting, finding pythons is largely a matter of being in the right place at the right time.

During the spring, radio-tagged male pythons searching for mates can lead biologists to other untagged animals.

Using "Judas snakes," biologists can sometimes locate breeding aggregations and capture several pythons at once.

both the female and one or more attending males. These radio-tracked snakes became known as *Judas snakes* because they "betrayed" the other snakes around them to python hunters. Some researchers and the media briefly touted the idea of using Judas snakes to control python populations, but they rarely mentioned the limitations of the method. If you are already conducting a radiotelemetry project, the Judas snake phenomenon may be an excellent way to find additional snakes for research purposes, but the expense and intensive effort involved in tracking pythons in South Florida render this method unrealistic for large-scale control. Additionally, because several males often attend a single female during the mating season, this method works best for finding male snakes and is totally ineffective for finding immature juveniles. Removal of large reproductive females, however, is more important for suppressing python populations.

DOGS

Most dogs have excellent chemoreception (sense of smell), and if properly trained can sniff out hidden animals of other species. Dogs have been used successfully to find desert tortoises and indigo snakes. They have also been used on Guam to find brown treesnakes, but primarily in airports and cargo areas rather than in the field. Detection in such areas has helped to prevent brown treesnakes from stowing away on

Dogs search cargo leaving Guam for stowaway brown treesnakes.

"Python Pete" was successful at finding pythons in experimental trials, but the effectiveness of using dogs to locate pythons in the Everglades has yet to be determined.

ships and airplanes and being introduced on other tropical islands such as Hawaii, where they would likely have devastating impacts on native bird populations.

Numerous people have proposed using dogs to detect pythons in Everglades National Park and its environs. National Park Service biologist Lori Oberhofer trained a beagle to detect pythons in 2005. Although "Python Pete" was able to find pythons during experimental trials, using him to find free-ranging pythons was impractical. First, a dog released into the Everglades faces numerous dangers. In addition to large pythons that could easily kill and consume a small dog, alligators and venomous snakes are common throughout the Everglades. Further, the rough limestone substrate can be extremely hard on the pads of dogs' feet. Finally, much of the Everglades is covered in shallow water most of the time, and even the best-trained dog would have difficulty tracking scents of free-ranging pythons there.

Dogs have been recorded in the diet of wild pythons in Asia.

Personnel from several institutions and agencies met in Everglades National Park in 2010 to discuss the use of dogs to detect pythons. All agreed that controlling the entire python population with dogs was too impractical to even discuss seriously, but that dogs might be effective in detecting pythons in certain areas, and as of 2011, Christina Romagosa and colleagues (Auburn University) are conducting trials to test the effectiveness of using trained dogs to detect pythons both in Everglades National Park and surrounding areas.

BOUNTY HUNTERS

The use of "bounty hunters" to eradicate, or at least control, python populations has received a lot of support from the media. The idea is to get local hunters and others who spend a lot of time outdoors to search for and capture or kill as many pythons as possible. State politicians instituted a program in 2009 allowing volunteers to hunt and kill pythons, although the hunters received no bounty payments. Unfortunately, the problem that plagues scientists looking for pythons plagues volunteer hunters as well—pythons are extremely hard to find. In the first year of the new program, volunteers found fewer than 40 pythons. Although some of the politicians who supported the program proclaimed it a huge success, the removal of 40 animals from a population that may very well number in the hundreds of thousands is likely inconsequential. Additionally, none of the snakes found by the volunteers went to scientists for examination. Future efforts should ensure that captured or killed pythons are given to scientists along with carefully collected data on the circumstances in which they were found.

Though widely touted as a success, a state-sanctioned python-hunting program yielded fewer than 40 pythons in 2009 while biologists and amateur snake hunters captured several hundred.

Bounty-type programs clearly will have little success for large-scale suppression or eradication of pythons, and they have the potential to cause other problems. First, uninformed hunters might kill nontarget species of snakes, including imperiled species such as eastern indigo snakes. Second, hunting is not allowed in national parks. Instituting a bounty program for pythons in Everglades National Park would set a bad precedent and likely result in considerable problems in the future. Regardless, amateur snake hunters have played an important role in finding and learning about pythons in Florida and will likely continue to provide important contributions to understanding more about invasive pythons and the impacts they will have.

Baited traps have captured several pythons in Key Largo.

Drift fences direct animals into traps.

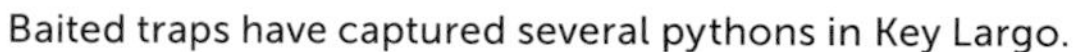

TRAPPING

Herpetologists often use snake traps, which generally consist of some sort of funnel trap into which snakes crawl and cannot find their way out. Traps can be baited (often with a live mouse or rat) and are frequently used in conjunction with a drift fence, a barrier constructed to divert animals moving through the environment toward the trap.

Because pythons are primarily ambush predators, they may be difficult to capture in traps.

Several factors would limit the utility of trapping pythons in South Florida as a control method. First, constructing drift fences in the hard limestone present in most areas would be very difficult. Drift fences are generally sunk several inches into the ground to prevent snakes from crawling under them. Second, snakes have to move in order to be trapped. Pythons do move, but probably less than some other species of snakes because they are primarily ambush hunters rather than active foragers. Finally, traps large enough to hold an adult python are expensive to build, and such large traps would probably catch many nontarget species of animals. Even if these factors could be overcome, the number of traps required to suppress populations or eliminate pythons from any area would probably be astronomical. Scientists studying methods to remove the venomous habu viper from a small village in Japan found that 494 traps per acre would be needed. Everglades

National Park encompasses nearly 1.4 million acres. Trapping such an area with the same intensity as that required for eliminating the habu would mean deploying and regularly checking more than 691 million traps! Aside from the impracticality of such an undertaking, it would likely have devastating impacts on native fauna.

Like some of the other control methods we have discussed, trapping on a smaller scale might be effective for removing pythons from critical areas. Although scientists working with Bob Reed, Kristen Hart, Gordon Rodda, members of Frank Mazzotti's lab, and Skip Snow made intensive efforts to trap pythons, as of 2010 they had captured only about half a dozen. USGS scientists are using traps on Key Largo to intercept pythons as they arrive on the island to help protect the endangered Key Largo woodrat. Trapping might also be effective in suppressing populations of pythons in other ecologically important areas such as bird rookeries and could be used in conjunction with barriers to prevent pythons from entering residential areas or from reaching the southern Florida Keys.

BIOLOGICAL CONTROLS

Scientists have been discussing potential methods of biological control that would either kill Burmese pythons or eliminate their ability to reproduce ever since it became apparent that the snakes were established in South Florida. Generally, biological controls, or *biocontrols*, are some sort of bacteria, virus, or parasite that specifically target only the invasive species. Biocontrols can be existing organisms or genetically engineered viruses or bacteria. Biocontrols have an advantage that the other methods of control

Baited traps have been used to capture pythons in a variety of habitats but with a low rate of success.

discussed above lack: they do not rely on humans being able to detect pythons. Although the idea is certainly appealing, effective biocontrols are time-consuming and expensive to develop, and their implementation can be problematic. If the disease-causing organism (such as a bacteria or virus) were to affect native species, the results could be catastrophic. Biocontrol possibilities for pythons do exist, however. A disease known as inclusion body disease that is fatal and apparently occurs only in boas and pythons warrants examination as a possible biocontrol agent.

Even a biocontrol agent that worked effectively only against its target species might not solve the python problem. Pythons could evolve resistance to the biocontrol. A similar situation occurred in Australia, where a virus was used as a biocontrol agent to attempt to control invasive European rabbits that were causing extensive environmental damage. Although the biocontrol agent was initially effective, rabbits resistant to the disease survived and reproduced, producing a resistant rabbit population that was unaffected by the virus. Releasing a biological control into the environment is essentially intentionally releasing one invasive species to control another. Because of the potentially devastating effect of biocontrols on natural ecosystems, such measures should be instituted only after careful deliberation.

REGULATIONS AND ENFORCEMENT

At present, few regulations limit citizens' ability to keep exotic animals in the United States. In many areas it is perfectly legal for anyone to own a full-grown tiger or 15-foot king cobra, despite the dangers such animals pose both to their owners and to other people. Governments—local and state—should be well aware of the dangers posed by some exotic pets. In 1994, for instance, George Dalrymple of Florida International University warned in a state-mandated report on Florida's invasive species that "the most significant potential problem in the next twenty years has to do with a category of non-indigenous species that is increasingly found as accidental, or intentional, releases in urban and suburban settings: the large constricting snakes (pythons and boas), and large predaceous monitor lizards." He went on to recommend "some means of regulating their sale and maintenance." Unfortunately, effective and reasonable regulations are rarely instituted proactively. Such has been the case with invasive Burmese pythons in South Florida. Regulations on the sale or keeping of large pythons might have prevented the introduction of pythons in Florida, but at the time of the introduction

Numerous efforts are under way to try to reduce the release of unwanted pet snakes into the wild.

no laws restricted keeping or selling such snakes. In fact, as of 2010, most states still had no laws preventing the release of exotic animals into natural environments. Soon after it became apparent that pythons were reproducing in South Florida, state and local agencies began encouraging pet owners not to release unwanted pets into the wild. By then, however, it was too late.

Florida state agencies have implemented a variety of campaigns to try to curb the release of exotic snakes. In January 2008 the Florida Fish and Wildlife Conservation Commission instituted regulations requiring owners of boas and pythons greater than 2 inches (5 cm) in diameter to purchase a $100 permit and have a microchip or PIT tag implanted in their snakes for identification purposes. In 2009 the state of Florida began offering "amnesty days," in which owners of unwanted pythons and other exotic pets could turn them in to authorities with no questions asked. In June 2010 a law banning personal possession of Nile monitors and seven species of large constricting snakes, including Burmese pythons, went into effect in Florida. People already properly licensed to own pythons were allowed to keep their snakes but were required to maintain a valid Reptile of Concern license for the remainder of the animal's life. Such laws may help to prevent the introduction and establishment of other species or of Burmese pythons in areas outside South Florida, but some people fear that pet owners who do not want to pay the $100 permit fee will release their pets and compound the problem. Enforcement of the law is difficult, and state agencies often do not have the personnel or resources to effectively regulate ownership of large constricting snakes.

Nearly two-thirds of the lizard species found in the eastern United States are exotic species established in South Florida.

As of September 2010 the U.S. Fish and Wildlife Service was considering including several species of large constricting snakes as "injurious wildlife" under the Lacey Act, partly on the basis of the 2009 report by Reed and Rodda. Such a listing would essentially prevent importation and interstate trade of Burmese pythons and other large constrictors and substantially reduce their numbers in the pet industry. The report has both supporters and opponents. Those with a vested interest in importing, breeding, and selling pythons are outraged and represent a formidable opposition. Python breeders and others in the reptile pet industry have launched aggressive campaigns against listing large constrictors under the Lacey Act, and numerous people on both sides of the debate have testified in Washington, D.C., about the matter.

Laws regulating the trade of exotic animals should be carefully considered, of course, in order to avoid knee-jerk reactions that result in unreasonable restrictions. And although laws regulating importation

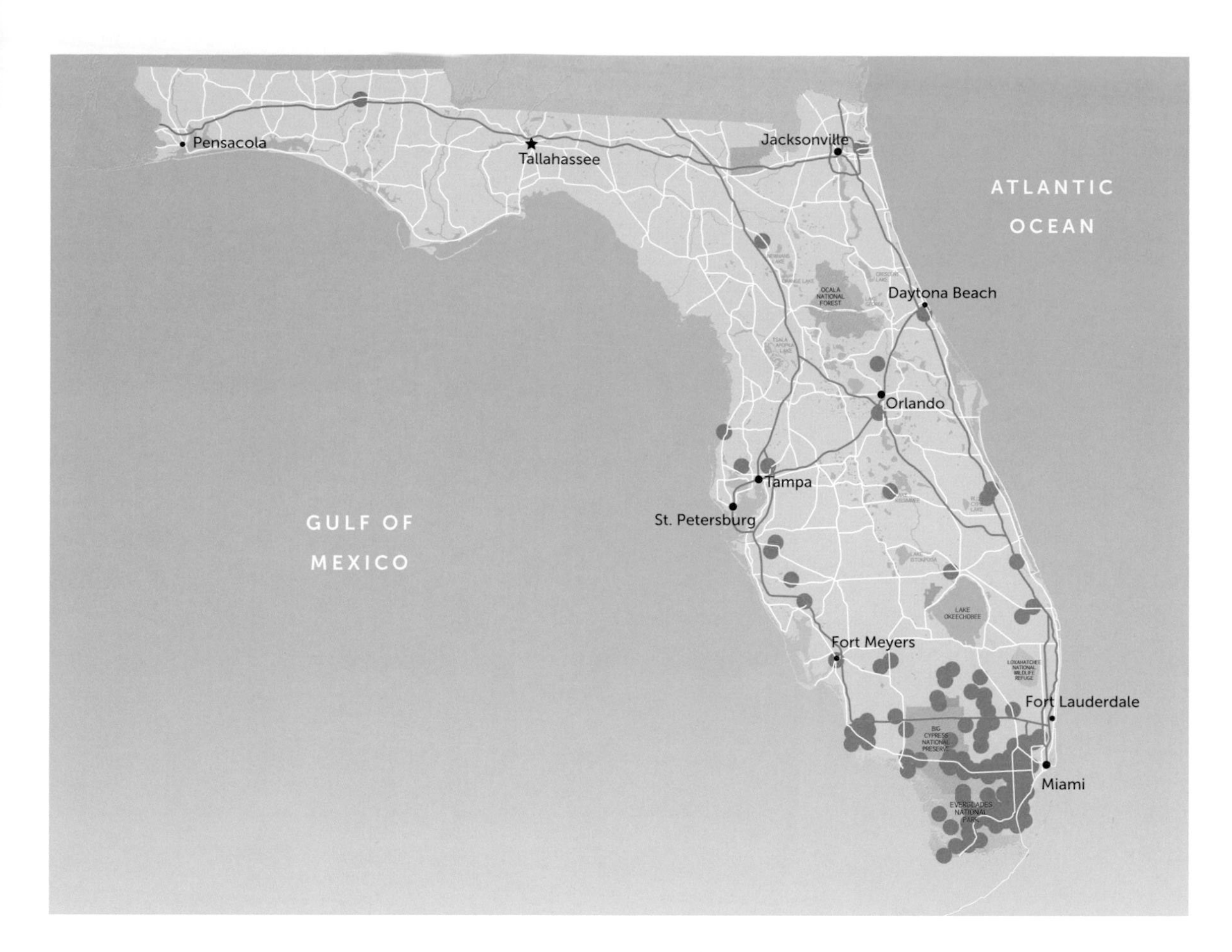

Localities of Burmese pythons found in Florida and reported to the Cooperative Invasive Species Management Area Website (www.evergladescisma.org) as of September 2010. Records in central and northern Florida probably represent released or escaped pets and are not part of established breeding populations. Adapted from ECISMA.

and sale of some species are certainly warranted, laws banning trade of pythons will not necessarily prevent future introductions.

WHAT IF YOU SEE A PYTHON?

Intentionally freed or escaped pythons and other large constricting snakes show up from time to time in various parts of the country. Any python found outside South Florida almost surely was once someone's pet. If you find a snake you suspect is a python, first make sure to correctly identify the species. Pythons are relatively easy to identify based on their size and color pattern. Any snake longer than 8 feet is probably not a native species. If a python is found, we do not recommend killing it. It may be someone's escaped pet, or you may have misidentified a native species. Call the proper authorities and ask to have the snake removed. The correct agency for such a situation will vary from place to place, and many animal control departments will not deal with snake calls. Contacting your state wildlife agency (in Florida, the Florida Fish and Wildlife Conservation Commission, www.myfwc.com) is usually

Many native snake species, such as this brown watersnake (right), have brown-blotched patterns and might be mistaken for a young python (below).

A team of "rapid responders" has been organized to investigate python reports in South Florida.

the best approach. In addition, nearly all states and many large cities have active herpetological societies with experts able and willing to assist with problem snakes. They may even be able to find a home for the animal. If you see a python in Everglades National Park or in its vicinity, you can call the python hotline that was established for just such situations. A simple Web search for "Everglades python hotline" should yield the proper phone number. You should be prepared to give an exact location and time of sighting for the python. If you have the opportunity, photograph the snake so that its identification can be confirmed. South Florida residents can also report sightings of invasive species to the Everglades Cooperative Invasive Species Management Area Web site (www.evergladescisma.org). This site provides useful information about invasive species and allows users to report sightings. An automated system on this Web site generates maps that document sightings of invasive species.

6 Pythons in the Pet and Skin Trades

Exotic reptiles, especially snakes, have always fascinated collectors, and international trade in snakes has a long history. The United States, Germany, and Japan have traditionally been the leading importers of live reptiles, and large-scale exotic animal importers have been doing business in these countries for decades. Historically, importers received animals from dealers around the world who had paid local villagers miniscule fees to collect all manner of local animals, including snakes. Animals were typically shipped in massive mixed-species shipments without special handling or care. Dealers seldom knew exactly what species they were ordering, and animals often arrived sick, dehydrated, or even dead. The animals were usually held in warehouses prior to their sale. Many animals died, escaped, or were released if they could not be sold. Animals coming into the United States often went to importers in or near Miami, Florida, and these locations became hotbeds for the introduction of exotic species. Although pet reptiles are being bred in captivity in ever-increasing numbers, large-scale reptile importation still occurs.

Bryan Christy's book *The Lizard King* recounts the interesting and sometimes nefarious history of the exotic reptile trade in the United States. The trade in exotic reptiles exploded in the 1970s and 1980s, spurred in part by the increase in global trade and the entrepreneurship of several large-scale reptile dealers. Snakes, lizards, and turtles became popular pets among a growing group of hobbyists and wealthy aficionados who viewed exotic animals, especially rare or dangerous ones, as status symbols. Over the course of a few decades, giant constrictors went from being relatively rare zoo specimens or curiosities to a mainstay of the pet industry.

Many pythons are sold each year at massive exotic reptile expositions held in different parts of the country.

BURMESE PYTHONS IN THE PET TRADE

Because of their size, beauty, generally calm temperament, and availability as imports, Burmese pythons have long been popular in the exotic snake trade; among the boas and pythons, only the ball python and boa constrictor surpassed them in popularity. Inexpensive, wild-caught juvenile Burmese pythons were available through several reptile dealers in the United States as early as the 1970s. According to records compiled by Bob Reed and Gordon Rodda (2009) the volume of trade in Burmese pythons fluctuated widely over the ensuing three decades, peaking in 1998 and 2000 when more than 20,000 were imported. Although early imports may have originated from throughout the species' range, harvesting of both subspecies of the Indian python (*P. m. molurus* and *P. m. bivittatus*) was curtailed in several south Asian countries (including India and Thailand) when the snakes received legal protection in the late 1970s and early 1980s. Since that time most imported Burmese pythons have presumably come from Southeast Asia (Vietnam, Cambodia, and Laos). Accurately assigning the origin of imported snakes is seldom possible, however, and considerable unreported shuffling of wild-caught pythons certainly occurs prior to their export. For example, the years just after Burmese pythons received protection in Thailand saw a spike of exports from Malaysia, where the species does not even occur. Presumably this spike represented snakes illegally harvested in Thailand and routed through Malaysia. Regardless of their exact country of origin, however, it is likely that most Burmese pythons imported since 1990 originated in tropical regions of Southeast Asia.

Although the trade in wild-caught Burmese pythons was already flourishing in the mid-1970s, the industry really took off after an amelanistic ("albino") individual was discovered in the late 1970s. This beautiful snake, which lacked all dark pigment and was instead spangled in lemon yellow, white, and pink, appeared in *National Geographic* magazine in 1981. Tom Crutchfield, a well-known South Florida reptile dealer, reportedly purchased it and two other pythons from a Thai dealer for $21,000 and brought them to the United States. Crutchfield then partnered with Bob Clark, one of the leaders in mass production of captive-bred pythons, who produced the first captive-bred albino hatchlings several years later. These albinos initially commanded astronomical prices—up to several thousand dollars apiece for hatchlings—spawning a new generation

NUMBERS OF BURMESE PYTHONS IMPORTED INTO THE UNITED STATES BETWEEN 1977 AND 2007

Data are from the Convention on International Trade in Endangered Species of Wild Fauna and Flora (CITES).

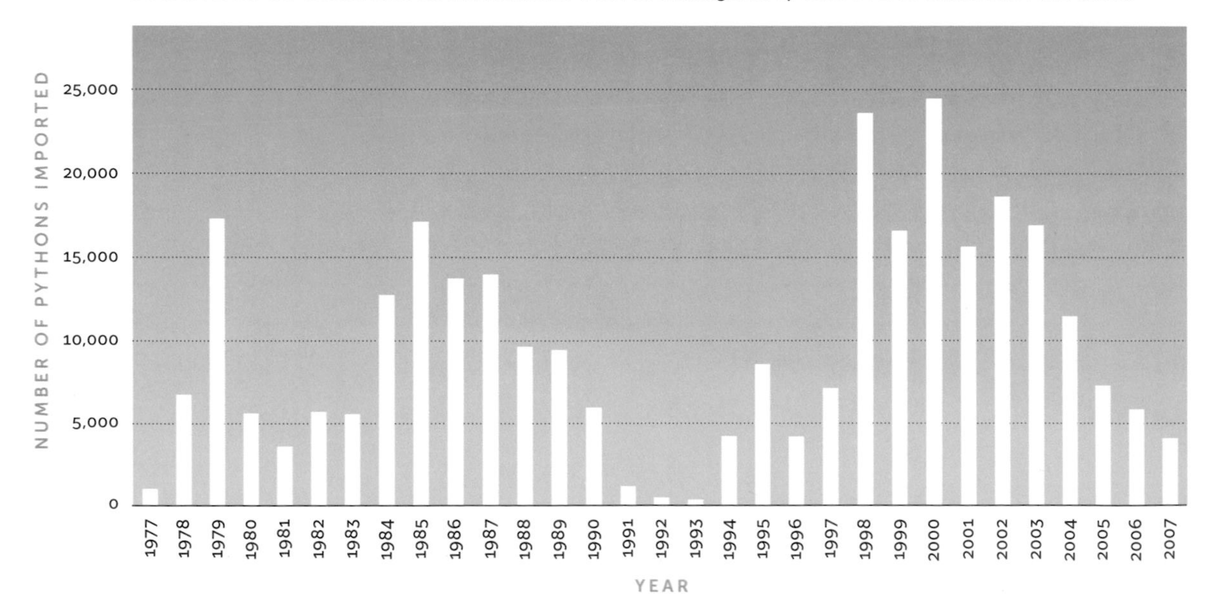

Selective breeding of captive Burmese pythons has produced a variety of attractive color morphs: labrinth (top left), green (top right), and albino (bottom).

The first albino Burmese python to enter the international pet trade appeared in a 1981 issue of *National Geographic* magazine.

of snake breeders who viewed their snakes as investments rather than simply as pets. Moreover, this event began a continuing quest among reptile dealers to find and propagate new python color morphs. Breeders carefully cultivated lineages of "designer" pythons, which now include "green," "patternless," "granite," "labyrinth," and "caramel" forms. Because these color morphs can be produced only through selective captive breeding of rare aberrantly patterned individuals, hobbyists have perfected techniques for quickly raising pythons to maturity and stimulating them to breed by manipulating temperature, lighting, food, and the presence of mates. Even though captive-bred Burmese pythons, including both normally patterned and designer morphs, are now available and generally inexpensive, significant numbers of wild-caught juvenile pythons continue to be imported from Southeast Asia.

RESPONSIBLE PET OWNERSHIP

The goal of this book is to discuss the status and biology of invasive pythons in the United States, not to provide a manual on care of captive pythons. Many excellent publications are available on that subject (see Further Reading), and we refer readers to those sources for detailed information on python care. We nevertheless wish to outline factors that prospective python owners should consider before purchasing a pet python in the hope of minimizing the introduction of additional pythons into the wild.

Before purchasing any pet, prospective owners must consider whether they are prepared to provide an appropriate environment and diet for the animal throughout its expected lifetime. Although appealing for their size and beautiful coloration, all of the giant constricting snakes discussed in this book can live more than 20 years and require large, expensive, escape-proof housing. The largest species, including Burmese pythons, reticulated pythons, African rock pythons, and anacondas, can reach lengths greater than 20 feet (6.1 m) and can outweigh an adult human. The sheer size and strength of these animals make them appropriate only for advanced hobbyists with extensive experience in keeping reptiles. Additionally, mature individuals need to eat large prey, such as chickens, pigs, or rabbits. Few casual owners are prepared to provide such expensive meals for many years. Finally, while many large constrictors, especially Burmese pythons, are quite docile when raised in captivity, they are very powerful animals, and no adult python should be handled or fed without at least two people present.

After careful consideration, many prospective snake owners decide that a smaller snake species is a more appropriate pet than one of the giants. Those who still wish to own a boa or python can choose from among many smaller, more manageable species that are commonly bred in captivity and are widely available, including ball pythons, rainbow boas,

Children's pythons, and blood pythons. Perhaps even more appealing are species of snakes native to the United States, which presumably pose less of a threat as potential invasive species. Commonly kept native snakes that are relatively easy to care for include rosy boas, corn snakes, rat snakes, and kingsnakes.

Once prospective owners have decided which species of snake to keep, they must first check state and local regulations to be sure that the species can be legally kept in their area and to determine if any permits are required. Several states have recently established permitting systems or restricted ownership of large constrictors, and Florida banned in-state sales of them in 2010. All of the species mentioned above are regularly bred in captivity, and numerous reputable dealers sell them. Captive-born snakes are a good choice for several reasons. In addition to not encouraging harvest from wild snake populations, captive-born snakes usually eat more readily, are healthier, and show superior coloration and temperament to wild-caught individuals.

Once purchased, the snake should be kept in a secure cage; many snakes are far better escape artists than their owners initially realize. Giant constrictors require a substantial amount of space but should never be allowed to range freely outside a cage. Free-ranging pythons could endanger other pets or people, including their owners, and are

Few owners have the space or resources necessary to properly care for a full-grown python (top). Many snake species are attractive alternatives to large pythons. Rosy boas (bottom) are attractive, small, and are bred in captivity in large numbers.

Each year, thousands of reticulated pythons are harvested for their skins and processed in slaughterhouses like this one (left) in Malaysia. Python skins are made into a variety of clothing and accessories (right), such as these reticulated python skin boots owned by one of the authors.

at high risk of escaping. Cages for large constrictors should be kept locked at all times to prevent escape or accident. The death of a 2-year-old girl who was killed by a pet Burmese python that escaped from its cage in Florida in 2009 is a stark reminder of the potentially tragic consequences of insecure housing.

Finally, unwanted snakes should never be released into the wild. Releasing pets is illegal in many areas, and the ecological and economic damage associated with establishment of exotic species can be catastrophic. Many pet owners find it surprisingly difficult to find a new home for an unwanted adult python. Most zoos quickly reach their capacity for keeping such snakes, and many pet adoption agencies refuse to accept giant constrictors. Inability to find a suitable new home, however, should never be considered an acceptable justification for releasing a pet python into the wild. Some states, such as Florida, have established "amnesty days" when unwanted pets—even those kept illegally—can be turned in to authorities without penalty to the owner. As a last resort, if a suitable new owner cannot be found, an unwanted python should be euthanized by a qualified veterinarian.

PYTHONS IN THE SKIN TRADE

In addition to collection of live animals for the pet trade, pythons are also harvested in large numbers in Southeast Asia for their skins and meat, and for their internal organs (particularly gall bladders), which are used in some forms of traditional medicine. Their large size and spectacular patterns make pythons attractive to clothing manufacturers worldwide, and millions of python skins have been exported from Southeast Asia since the 1930s. Most of the skins were used in the manufacture of footwear, bags, and belts, with the principal markets in Europe and the United States.

The commercial trade in python hides is generally centered in Southeast Asia, and the reticulated python, Burmese python, and blood python are the most commonly harvested species. A comprehensive report prepared by the Convention on International Trade in Endangered Species of Wild Fauna and Flora

(CITES) notes that 22,000 to 189,000 Burmese python skins were imported from Southeast Asia each year between 1980 and 1987. These figures are almost certainly underestimates, however, because they do not include cargo recorded by length or weight. Most Burmese python skins traded in the early to mid-1980s apparently originated in Thailand, and that country reported a high of 188,000 skins exported in 1985—the year before the species was given legal protection; exports from that country showed a steep decline thereafter. The subsequent sharp peak in python exports from Malaysia (where they do not occur) indicates considerable unreported rerouting of skins prior to export, complicating our ability to accurately assess the volume and origin of python skins. Overall, trade in Burmese python skins declined following a peak in the mid-1980s and probably continues to decline today.

In the 1980s between 200,000 and 700,000 reticulated pythons were killed for their skins each year.

Although the recorded numbers of Burmese pythons harvested in the 1980s may seem enormous, they are dwarfed by the harvest of reticulated pythons. Export records indicate that 200,000 to 700,000 reticulated pythons were harvested annually during that decade, primarily in Indonesia. Although the trade in reticulated python skins also peaked in the mid-1980s, large-scale harvest continues into the 21st century. Because reticulated pythons are the mainstay of the python skin trade, much of what we know about the attributes of harvested snakes and the effects of harvesting on python populations derives from studies of this species. University of Sydney biologist Rick Shine visited python slaughterhouses in Sumatra in the 1990s and examined more than 1,800 reticulated pythons that were harvested for their skins, the largest sample of wild-caught giant Asian pythons ever examined by scientists (see Shine 1991, 1999). He found that most of the reticulated pythons harvested in northern Sumatra were taken from agricultural areas and around villages, where they fed extensively on rats and domestic chickens. Although most of the pythons in the sample were mature, the average size was relatively small (less than 10 feet [3 m]). Pythons from less developed areas in southern Sumatra were larger and ate a wider range of prey. Shine's studies suggest that reticulated pythons can withstand substantial harvesting and persist in developing areas by feeding on rats, growing rapidly, and producing large numbers of offspring. These attributes, which Burmese pythons share, may hamper control efforts in Florida, and it is likely that we will have to remove very large numbers of them to have an appreciable impact on population density. Establishing a market for skins and meat of Florida pythons might stimulate harvesting and help control the invasive population but seems unlikely to be lucrative enough to subsidize professional python hunters.

7 Other Species at Risk of Becoming Established in the United States

Although this book focuses on invasive Burmese pythons, which are now firmly established in southern Florida, five other species or species groups of pythons and boas deserve mention as potential invasive species in Florida and elsewhere in the United States. Two of these species, the African rock python and boa constrictor, are currently established and reproducing in localized areas in southern Florida but have not expanded as explosively or become as widespread as the Burmese python. The reticulated python and ball python are both extremely popular in the pet trade and turn up regularly in South Florida and elsewhere, although they are not suspected to be established or reproducing in the wild. The green and yellow anacondas, which have been found in small numbers in Florida, have highly aquatic lifestyles that would presumably make them successful in the Everglades. Robert Reed and Gordon Rodda covered all of these species but the ball python in their 2009 report on giant constrictors, and we highly recommend that anyone interested in details of this issue review their report.

Although we restrict our discussion in this chapter to selected pythons and boas, several other snake species, some potentially dangerous to people or wildlife, are thought to be at high risk of becoming established in the United States as well. For example, a recent risk assessment study by Ikuko Fujisaki and colleagues (2009) suggested that the venomous African puff adder is at high risk of becoming established in South Florida.

CHARACTERISTICS OF SEVEN SPECIES OF PYTHONS AND BOAS THAT ARE EITHER ESTABLISHED OR ARE AT RISK OF BECOMING ESTABLISHED AS INTRODUCED SPECIES IN THE UNITED STATES

SPECIES	MAX SIZE	NATIVE RANGE	HABITAT	DIET	STATUS IN THE UNITED STATES
Burmese python	27 ft.	Tropical and temperate southern and southeastern Asia	Generalist	Primarily mammals and birds	Well established in a large area of South Florida.
African rock python	23 ft.	Sub-Saharan Africa	Generalist	Primarily mammals and birds	Established in a restricted area of South Florida.
Boa constrictor	15 ft.	Northern Mexico to Argentina	Generalist	Mammals, birds, and large lizards	Established in a restricted area of South Florida.
Green anaconda	26 ft.	Amazon Basin	Tropical aquatic	Mammals, birds, large lizards, crocodilians, snakes, and fish	Few have been found; not known to be established.
Yellow anaconda	14 ft.	Southern Amazon Basin and Argentina	Tropical/temperate aquatic	Mammals, birds, large lizards, crocodilians, snakes, and fish	Few have been found; not known to be established.
Reticulated python	33 ft.	Tropical Southeast Asia	Tropical generalist	Primarily mammals and birds	Few have been found; not known to be established.
Ball python	6 ft.	Central Africa	Tropical generalist	Primarily small mammals	Moderate numbers found; not known to be established.

A small population of African rock pythons is established just northeast of Everglades National Park.

AFRICAN ROCK PYTHONS

Most experts consider Africa's largest constrictor, the rock python, to comprise two closely related species: the northern African python (*Python sebae*) and the southern African python (*Python natalensis*). Together these two species range across much of sub-Saharan Africa with the exception of very high elevations and extreme desert environments. They are close relatives of the Indian python and are similar to it in both appearance and biology. African pythons inhabit a wide range of habitats in their native range but are frequently associated with water and often use mammal burrows for shelter. Like Indian pythons, they prey primarily on mammals and birds, and females may reach 20 feet (6.1 m) in length. There is evidence that male African pythons may grow larger than male Indian pythons, but females are certainly larger in both species. Although African pythons have a reputation for being vicious or bad-tempered when confronted or kept in captivity, this probably does not translate into a propensity to attack humans in a predatory fashion. Only a limited number of substantiated predatory attacks by African pythons on humans have been reported in their native range. Like other pythons, African pythons lay large clutches of eggs (average 30–50, up to 100), which the female guards until hatching. Most scientists believe that they do not shiver to warm their eggs as Indian pythons do.

Because of their somewhat more subdued coloration and bad disposition, African pythons are much less common in the pet industry than the large Asian pythons or the boa constrictor. African pythons have been imported into the United States in fairly large numbers, however, with imports averaging about 1,000 per year according to CITES records. Because they are much less common as pets, escaped or released African pythons are rare in the United States. Several reports have occurred from a relatively small area west of Miami and adjacent to the Everglades, and African pythons are clearly established there (see Reed and colleagues 2010). Five African pythons were recovered from this area in 2009, including two hatchlings, suggesting that reproduction was occurring. Prompted by this troubling observation, concerned citizens and scientists from the USGS, National Park Service, Florida Fish and Wildlife Conservation Commission, South Florida Water Management District, and other agencies organized African python searches in this area in 2010. Between January and June 2010 these searches yielded at least 14 African python captures or sightings, including 2 gravid (pregnant) females and an old nest containing shells from 20 hatched eggs. Efforts are currently under way to combat this incipient population.

It is not surprising that African pythons have gained a foothold in southern Florida. This species is so similar ecologically to the Burmese python that it would be logical to assume that areas suitable for Burmese

Boas and pythons have legs! The tiny clawlike structures found on either side at the base of their tail are actually vestigial limbs.

Pregnant females (left) and shells of hatched eggs (right) are firm evidence that African pythons are reproducing in at least one location in Florida.

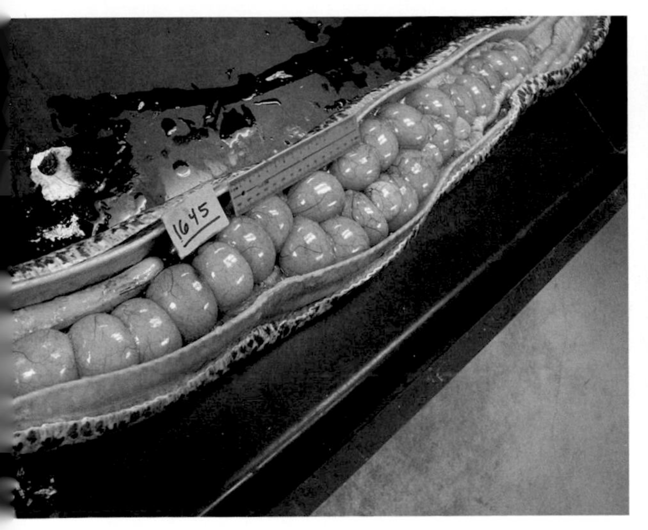

pythons would also be suitable for African pythons. Indeed, prior to realization that African pythons were successfully established, Ikuko Fujisaki and colleagues (see Further Reading) predicted that they were among four snake species at highest risk of becoming established in South Florida. Although the discovery of a reproducing population of African pythons prompted headlines about "killer snakes," we have no reason to believe that African pythons will be any more harmful to native ecosystems or to people than Burmese pythons. In fact, the area where African pythons have been found is within or adjacent to the introduced range of Burmese pythons, and the two species may compete with each other for prey or habitat. Media coverage of African pythons has interpreted the irritable disposition of captive African pythons as evidence that they are more vicious, and therefore more dangerous to people and wildlife than Burmese pythons. It is important to remember that the "nasty" disposition of African pythons represents a defensive response to a perceived predator (humans), however, and probably does not reflect predatory behavior. In fact, most wild Burmese pythons in the Everglades also do not hesitate to hiss, bite, and emit musk when captured. Another myth about African pythons asserts that they will interbreed with Burmese pythons, producing "killer hybrid snakes" that are more vicious and dangerous than either of the two parent species.

An intensive search of the area occupied by introduced African pythons in January 2010 yielded five pythons and a nest that probably hatched the preceding summer.

Most introduced African pythons have been found in disturbed habitats such as this large pile of melaluca logs and debris.

LOCALITIES OF SMALL ESTABLISHED POPULATIONS OF AFRICAN PYTHONS AND BOA CONSTRICTORS IN SOUTH FLORIDA

African rock python   Boa constrictor

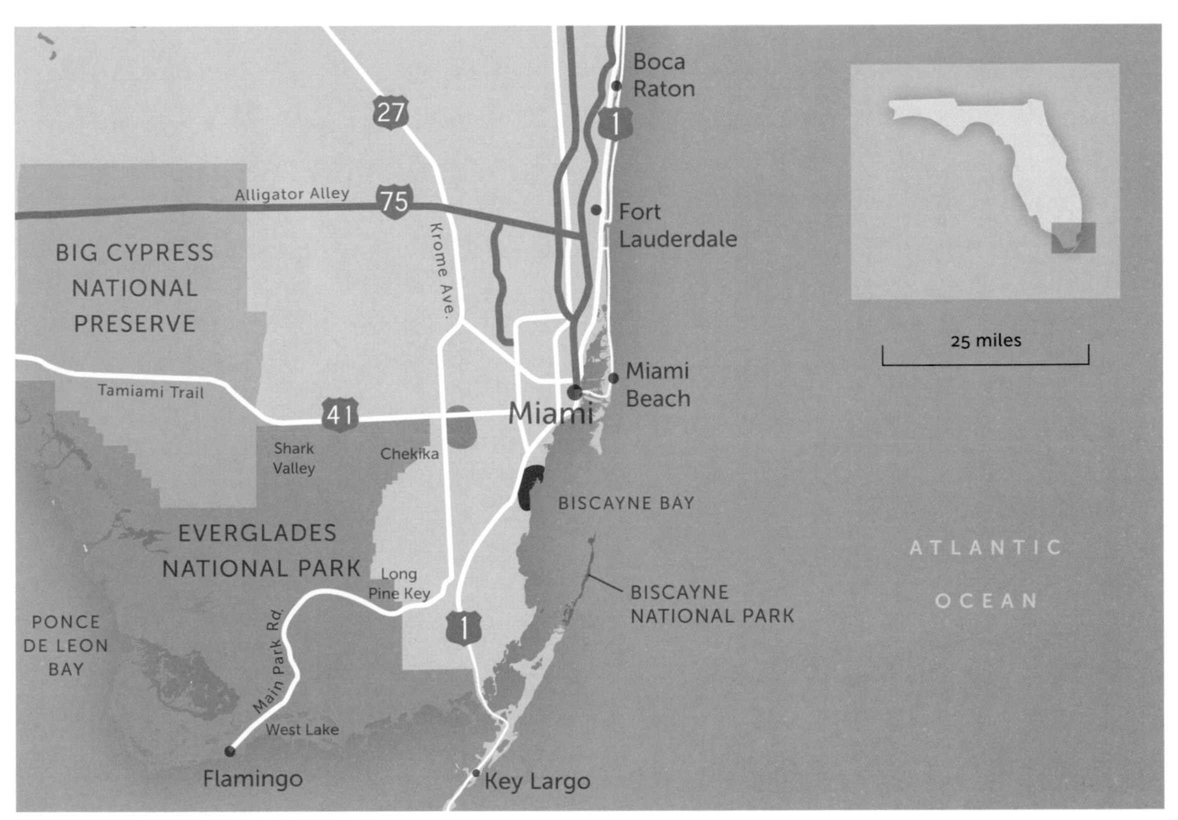

Although these two species have been successfully crossbred in captivity, we have no evidence of hybridization in the wild or any reason to suspect that a hybrid between the two would be anything other than a snake with a color pattern somewhere in between those of its parents. Moreover, biologists are optimistic that effective control measures might be able to eliminate African pythons or at least confine them to a small area. The Burmese python population was not recognized as being established until it had already spread over a vast and inaccessible area, but the African python population in South Florida is apparently still fairly small and restricted to a limited and fairly accessible area.

BOA CONSTRICTOR

The boa constrictor (*Boa constrictor*) has been extremely popular in the pet industry for decades. The species ranges widely from northern Mexico to central Argentina and occurs in many different habitats, including tropical rainforest, dry desert scrub, and the temperate grassland of the Argentine Pampas. Numerous subspecies exhibiting various color patterns are recognized, but those most common in the pet industry are from Colombia and Central America and are patterned with dark brown or black hourglass-shaped saddles on a tan or brown background. Many individuals from South America have a bright chestnut or reddish tail, leading to the common name red-tailed boa. Boa constrictors are very large snakes, but they do not regularly reach the extreme proportions of the Old World giant pythons or the green anaconda. Although tales of very large boa constrictors abound, the largest verified individuals were not much over 14 feet (4.3 m) in length. Most wild individuals are considerably smaller than that, and several dwarfed populations inhabit islands off Central and South America.

Wild boa constrictors exhibit a wide range of behaviors. Those inhabiting rainforests are frequently found in trees, where they presumably ambush prey. Boas inhabiting drier climates may spend considerable time underground in burrows made by other animals, or even in caves. They eat a variety of prey including birds, small to mid-sized mammals, and large lizards such as iguanas and tegus. Although they are frequently found along rivers and swim well, boas are less aquatic than anacondas and some Old World pythons. Also unlike the giant pythons, boas give birth to live young and have smaller litter sizes, generally 20 to 65 young. Boas do not regularly attack humans, and we know of no documented predatory attacks by wild boas. A 9-foot (2.7 m) pet boa constrictor that killed its owner in Nebraska in 2010 was apparently the first to kill a person in the United States, and possibly the first anywhere.

Three factors contribute to the risk boa constrictors pose as invasive species in the United States: (1) their popularity in the pet trade, (2) the wide range of climates and habitats they inhabit in their native range, and (3) their proven track record as an invasive species. The boa constrictor is second only to the ball python in popularity, and CITES records indicate that more than 150,000 boa constrictors were imported into the United States between 1983 and 2000. Like Burmese, reticulated, and ball pythons, boas are now

A population of introduced boa constrictors has existed in a small section of South Miami since at least 1990.

An introduced population of boa constrictors also exists on Aruba.

frequently bred in captivity, but substantial importation of wild-caught individuals continues. In accordance with their popularity, escaped or released boa constrictors are found with some regularity, especially in Florida.

The wide range of habitats boa constrictors occupy in their native range suggests that they are very adaptable, and indeed, there are at least three examples of successful establishment of boa constrictors outside their native range. These include established and growing populations on the islands of Cozumel and Aruba and an introduced population in Florida around the Deering Estate at Cutler, a Miami-Dade County park property. An established population may also exist in Puerto Rico. The original source of the boa constrictor population at the Deering Estate is unknown, but more than 100 boas have been found there since the population was discovered in 1989. Most were young snakes, and many were captured during a single year, 1996, when at least two female boas apparently gave birth on the property. The difficulties associated with differentiating wild boas from escaped pets make it hard to determine whether this population is expanding beyond the property. Although many individual boa constrictors have been found in the Everglades and elsewhere in South Florida, we know of no other established populations on the U.S. mainland. When a large boa was captured on Big Pine Key in 2010, Kenneth Krysko of the University of Florida was able to match its pattern to photos from previous reports, demonstrating that this individual had survived in the wild for at least 2 years. Perhaps fortunately, the great majority of imported boa constrictors come from tropical South and Central America and may not be able to tolerate cooler temperatures in the United States outside southern Florida. Presumably, boa constrictors from more temperate regions of Argentina and Mexico would pose the greatest risk as invasive species, but boas from these areas are relatively rare in the pet trade.

Yellow anacondas inhabit much more temperate climates than green anacondas but like their larger relatives are highly aquatic.

ANACONDAS

The anacondas are a group of four closely related and highly aquatic species native to mainland South America. Like their relatives the boas, they give birth to live young. The green anaconda (*Eunectes marinus*) and the yellow anaconda (*Eunectes notaeus*) are widespread in South America and are occasionally kept as pets. The Beni anaconda (*Eunectes beniensis*) and DeSchauensee's anaconda (*Eunectes deschauenseei*) are poorly known, have restricted native ranges, and are seldom if ever kept as pets in the United States. Consequently, we restrict our discussion of anacondas here to the green and yellow anacondas.

The green anaconda, which is native to tropical regions of the Amazon Basin, is the largest of the anacondas. The body is olive green with dark circular blotches down its length. Although reticulated pythons of Southeast Asia probably grow slightly longer, green anacondas can reach lengths in excess of 25 feet (7.6 m) and are unquestionably the most massive snakes in the world. Wild green anacondas certainly top 200 pounds (91 kg), and well-fed captives have weighed in at more than 500 pounds (227 kg). The yellow anaconda is much smaller, only

Green anacondas (above) are restricted to tropical South America and are highly aquatic. This yellow anaconda (left) is one of two that have been found in Big Cypress National Preserve in recent years.

about 12 feet (3.7 m) long, and is yellowish with blotches that are more irregular than those of the green anaconda. The yellow anaconda is found in cooler climates than the green anaconda, inhabiting subtropical and temperate regions of western Brazil, Paraguay, Bolivia, and eastern Argentina. Both anacondas are highly aquatic and are never found far from water. They also apparently feed on a wider range of prey than most other large constrictors, and in addition to mammals and birds also frequently consume caimans, turtles, other snakes, and fish.

Although neither green nor yellow anacondas are nearly as common in the pet trade as ball pythons, Burmese pythons, reticulated pythons,

The green anaconda (top) is unquestionably the heaviest snake in the world. Large green anacondas can sometimes be found in drying wetlands (bottom).

Large constrictors such as this reticulated python are formidable animals.

and boa constrictors, they warrant concern as potential invasive species. Both would almost certainly feel at home in the vast subtropical wetlands of the Everglades. The green anaconda is the more popular of the two species, and many reptile keepers have dreamed of owning one at one time or another. Although few hobbyists have the proper facilities to house or feed such a massive semiaquatic reptile, several hundred green anacondas have been imported annually since the late 1970s, and escaped or released green anacondas are occasionally reported. A road-killed male green anaconda was found in 2004 on the Tamiami Trail (U.S. 41) in Fakahatchee Strand Preserve State Park in the western portion of the Everglades. Green anacondas are restricted to tropical climates in their native range, however, and are unlikely to become established in the United States outside southern Florida or extreme southern Texas, although they could threaten Hawaii or Puerto Rico.

Only advanced reptile hobbyists should attempt to keep large constrictors such as pythons and anacondas.

Yellow anacondas are relatively rare in the pet trade, and CITES reports fewer than 2,000 imported into the United States in recent decades. Because they can tolerate cooler climates than most other boas and pythons, however, yellow anacondas have the potential to become established in many areas of the southeastern United States that combine extensive aquatic habitats with relatively mild winter temperatures. Indeed, two yellow anacondas were found in or near Big Cypress National Preserve in 2007 and 2008. Both were probably recently released or escaped individuals, and no established population of either anaconda species is known. Because anacondas are so thoroughly aquatic, however, an introduced population could easily go undetected for years.

RETICULATED PYTHON

The reticulated python (*Python reticulatus*) deserves mention as a potential invasive species because of its size and its popularity in the pet industry. Reticulated pythons have been imported at the rate of at least several thousand per year since the late 1970s, and are now bred in captivity in large numbers. The reticulated python is probably the longest snake in the world, very likely reaching lengths approaching 30 feet (9.1 m) in the wild. It is usually fairly slender compared with other python species, however, and certainly does not attain the massive weights of large green anacondas. Reticulated pythons primarily inhabit tropical rainforests in Southeast Asia from Bangladesh to Vietnam and most of the Indo-Pacific islands west of New Guinea. Virtually everything we know about their biology stems from anecdotal reports from the colonial era or data collected from snakes harvested for the skin trade. Reticulated pythons apparently occur in a variety of habitats, especially those near water. Small individuals are often found in trees and are relatively common in some agricultural areas, including palm plantations and rice paddies. Like other pythons, reticulated pythons feed primarily on warm-blooded prey, and large individuals are probably capable of consuming most mammal and bird species found within their native range. Documented prey items include everything from small bats, tree shrews, and ducks to deer, porcupines, and even a sun bear. Several authors have suggested that large reticulated pythons feed preferentially on wild boar, but examination of individuals collected for the skin trade showed that adults eat a variety of large mammals and birds; smaller individuals collected from agricultural areas had fed predominantly on rats and domestic chickens. The reticulated python also probably deserves its reputa-

Their beautiful coloration and large size make reticulated pythons popular among reptile hobbyists.

The reticulated python is probably the longest snake in the world. This individual exceeded 22 feet (7 m) in length.

This large reticulated python was photographed crossing a road at night in Everglades National Park in 1997.

tion as the snake most likely to kill and consume a human. Numerous attacks have been reported, and several people, even adults, have been killed and eaten by reticulated pythons in the wild. Captive reticulated pythons have killed several pet owners in the United States.

Although likely untrue, there is one account of a python in Asia taking and swallowing the corpse during a funeral.

Although they are quite popular in the pet trade and certainly escape or are released with some frequency in the United States, the generally tropical distribution of reticulated pythons probably makes them less risky as invasive species than most of the other large constrictors. Because of their popularity as pets, escaped or released reticulated pythons are found fairly frequently, especially in Florida. In 1989 a massive 22-foot (6.7 m) reticulated python was found living under a house in Fort Lauderdale, and several reticulated pythons have been found in Everglades National Park. There is no indication that established populations of this species exist in the United States, and climate-matching studies have indicated that suitable climate for this species exists within the continental United States only in extreme southern Florida and possibly Texas. If reticulated pythons were to become established, however, they could pose a serious risk to native wildlife, domestic animals, and potentially to humans.

BALL PYTHON

The ball, or regal, python (*Python regius*) is one of the smallest pythons and is unquestionably the most popular pet python in the United States today. Native to central and western Africa, ball pythons reach only about 5 feet (1.5 m) in length but are typically very stout-bodied. They generally have a tan to gold ground coloration with irregular dark markings. As is the case with other popular python species, breeders have produced many color morphs, including albino, striped, and piebald. The newest morphs often sell for thousands of dollars. True to their name, ball pythons defend themselves by coiling into a tight ball with the head protected in the center. Ball pythons occur in a variety of habitats and are common in some agricultural areas of West Africa. They feed primarily on mammals, particularly rodents. Because they are so small, ball pythons pose little threat to humans or domestic animals.

At least nine people have been killed by their pet pythons in the United States since 1980.

Ball pythons are tremendously popular pets and have been, by a large margin, the most frequently imported and least expensive python species. Between 1989 and 2000 alone more than 350,000 ball pythons were imported into the United States from West Africa, particularly from Ghana, Togo, and Benin. Most were newly hatched young born to gravid females collected from the wild. Although ball pythons are now bred extensively in captivity, wild ones are still imported in large numbers. Because of their popularity as pets, ball pythons are probably the species that most frequently escapes from captivity. Escaped ball pythons are found regularly in most major metropolitan areas in the United States. We have found them near our respective homes in Charlotte, North Carolina, and Athens, Georgia. Although ball pythons are reported regularly from scattered locations in South Florida, as of 2010 no established or reproducing populations were known. Because this species flourishes in agricultural areas of its native range, however, it would presumably succeed in similar habitats in southern Florida, Texas, or California where severe winter freezes do not regularly occur. Because the native range of this species is primarily tropical, it is unlikely to pose much risk of becoming established in areas of the United States that receive regular winter freezes.

Designer piebald ball pythons initially sold for thousands of dollars.

This presumably escaped or released ball python was found on Key Largo.

Further Reading

The sources listed below are mentioned in the text or offer additional information about pythons and especially about invasive pythons in Florida. Their inclusion does not mean that we can verify the validity of the information they present or that we necessarily agree with all the positions they espouse.

Andreadis, P. T. Forthcoming. *Python molurus bivittatus* (Burmese python). Reproducing population. *Herpetological Review.*

Avery, M. L., R. M. Engeman, K. L. Keacher, J. S. Humphrey, W. E. Bruce, T. C. Mathies, and R. E. Mauldin. 2010. Cold weather and the potential range of introduced Burmese pythons. *Biological Invasions* 12:2955–2958.

Barker, D. G., and T. M. Barker. 1994. *Pythons of the World.* Vol. 1: *Australia.* Lakeside, Calif.: Advanced Vivarium Systems.

Barker, D. G., and T. M. Barker. 2008. A critique of the analysis used to predict the climate space of the Burmese python in the United States by Rodda et al. (2008, 2009) and Reed and Rodda (2009). *Bulletin of the Chicago Herpetological Society* 45:97–106.

Barker, D. G., and T. M. Barker. 2008. The distribution of the Burmese python, *Python molurus bivittatus. Bulletin of the Chicago Herpetological Society* 43:33–38.

Bhupathy, S., and V. S. Vijayan. 1989. Status, distribution and general ecology of the Indian python, *Python molurus molurus* Linn. in Keoladeo National Park, Bharatpur, Rajasthan. *Journal of the Bombay Natural History Society* 86:381–387.

Bilger, B. 2009. Swamp things. *New Yorker* 85:80–89.

Christy, B. 2008. *The Lizard King: True Crimes and Passions of the World's Greatest Reptile Smugglers.* New York: Twelve, Hatchette Book Group.

Dalrymple, G. H. 1994. Non-indigenous amphibians and reptiles. Pages 67-71, 73-78 in D. C. Schmitz and T. C. Brown, project directors, *An Assessment of Invasive Non-indigenous Species in Florida's Public Lands.* Technical Report TSS-94-100. Tallahassee: Florida Department of Environmental Protection.

de Vosjoli, P. 2005. *Burmese Pythons: Plus Reticulated Rythons and Related Species.* Irvine, Calif.: Advanced Vivarium Systems.

Dorcas, M. E., J. D. Willson, and J. W. Gibbons. 2011. Can invasive Burmese pythons inhabit temperate regions of the southeastern United States? *Biological Invasions* 13:793–802.

Dove, C. J., R. W. Snow, M. R. Rochford, and F. J. Mazzotti. 2011. Birds consumed by the invasive Burmese python (*Python molurus bivittatus*) in Everglades National Park, Florida, USA. *The Wilson Journal of Ornithology* 123:126–131.

Fujisaki, I., K. M. Hart, F. J. Mazzotti, K. G. Rice, S. Snow, and M. Rochford. 2009. Risk assessment of potential invasiveness of exotic reptiles imported to South Florida. *Biological Invasions* 12:2585–2596.

Gibbons, J. W., and M. E. Dorcas. 2005. *Snakes of the Southeast.* Athens: University of Georgia Press.

Greene, H. W. 1997. *Snakes: The Evolution of Mystery in Nature.* Los Angeles: University of California Press.

Groombridge, B., and R. Luxmoore. 1991. *Pythons in South-east Asia. A Review of Distribution, Status, and Trade in Three Selected Species.* Report to CITES Secretariat, Lausanne, Switzerland.

Groot, T. V. M., E. Bruins, and J. A. J. Breeuwer. 2003. Molecular genetic evidence for parthenogenesis in the Burmese python, *Python molurus bivittatus.* Heredity 90:130–135.

Harvey, R. G., M. L. Brien, M. S. Cherkiss, M. E. Dorcas, M. Rochford, R. W. Snow, and F. J. Mazzotti. 2008. Burmese pythons in South Florida: Scientific support for invasive species management. University of Florida IFAS Extension. At http://edis.ifas.ufl.edu/uw286.

Henderson, R. W., and R. Powell, eds. 2007. *Biology of Boas and Pythons.* Eagle Mountain, Utah: Eagle Mountain Publishing.

Holbrook, J., and T. Chesnes. 2011. An effect of Burmese pythons (*Python molurus bivittatus*) on mammal populations in southern Florida. *Florida Scientist* 74:17–24.

Jacobs, H. J., M. Auliya, and W. Böhme. 2009. Zur taxonomie des dunklen tigerpythons, *Python molurus bivittatus* Kuhl, 1820, speziell der population von Sulawesi. *Sauria* 31:5–16.

Kraus, F. 2009. *Alien Reptiles and Amphibians: A Scientific Compendium and Analysis.* New York: Springer.

Mattison, C. 1991. *A–Z of Snake Keeping.* New York: Sterling Publishing Company.

Mazzotti, F. J., M. S. Cherkiss, K. M. Hart, R. W. Snow, M. R. Rochford, M. E. Dorcas, and R. Reed. 2011. Cold-induced mortality of invasive Burmese pythons in South Florida. *Biological Invasions* 13:143–151.

Meshaka, W. E. Jr., B. P. Butterfield, and J. B. Hauge. 2004. *The Exotic Amphibians and Reptiles of Florida.* Malabar, Fla.: Krieger Publishing.

Meshaka, W. E. Jr., W. F. Loftus, and T. Steiner. 2000. The herpetofauna of Everglades National Park. *Florida Scientist* 63:84–103.

Murphy, J. C., and R. W. Henderson. 1997. *Tales of Giant Snakes: A Natural History of Anacondas and Pythons.* Malabar, Fla.: Krieger Publishing.

Pope, C. H. 1961. *The Giant Snakes.* New York: Alfred A. Knopf.

Pyron, R. A., F. T. Burbrink, and T. J. Guiher. 2008. Claims of potential expansion throughout the U.S. by invasive python species are contradicted by ecological niche models. PLOS ONE 3:e2931.

Reed, R. N. 2005. An ecological risk assessment of nonnative boas and pythons as potentially invasive species in the United States. *Risk Analysis* 25:753–766.

Reed, R. N., K. L. Krysko, R. W. Snow, and G. H. Rodda. 2010. Is the northern African rock python (*Python sebae*) established in southern Florida? *IRCF Reptiles and Amphibians* 17:52–54.

Reed, R. N., and G. H. Rodda. 2009. *Giant Constrictors: Biological and Management Profiles and an Establishment Risk Assessment for Nine Large Species of Pythons, Anacondas, and the Boa Constrictor.* U.S. Geological Survey Open-File Report 2009-1202.

Rodda, G. H., C. S. Jarnevich, and R. N. Reed. 2009. What parts of the U.S. mainland are climatically suitable for invasive alien pythons spreading from Everglades National Park? *Biological Invasions* 11:241–252.

Rodda, G. H., C. S. Jarnevich, and R. N. Reed. 2011. Challenges in identifying sites climatically matched to the native ranges of animal invaders. *PLoS ONE* 6:e14670.

Rodda, G. H., Y. Sawai, D. Chiszar, and H. Tanaka. 1999. *Problem Snake Management: The Habu and the Brown Treesnake*. Ithaca: Cornell University Press.

Secor, S. M., and J. Diamond. 1998. A vertebrate model of extreme physiological regulation. *Nature* 295:659–662.

Shine, R. 1991. *Australian Snakes: A Natural History*. Ithaca: Cornell University Press.

Shine, R. 1999. Reticulated pythons in Sumatra: Biology, harvesting and sustainability. *Biological Conservation* 87:349–357.

Smith, H. T., A. Sementelli, W. E. Meshaka, and R. M. Engeman. 2007. Reptilian pathogens of the Florida Everglades: The associated costs of Burmese pythons. *Endangered Species Update* 24:63–71.

Snow, R. W., M. L. Brien, M. S. Cherkiss, L. Wilkins, and F. J. Mazzotti. 2007. Dietary habits of the Burmese python, *Python molurus bivittatus*, in Everglades National Park, Florida. *Herpetological Bulletin* 101:5–7.

Snow, R. W., K. L. Krysko, K. M. Enge, L. Oberhofer, A. Warren-Bradley, and L. Wilkins. 2007. Introduced populations of *Boa constrictor* (Boidae) and *Python molurus bivittatus* (Pythonidae) in southern Florida. Pp. 416–438 in R. W. Henderson and R. Powell, eds., *Biology of the Boas and Pythons*. Eagle Mountain, Utah: Eagle Mountain Publishing.

Snow, R. W., A. J. Wolf, B. W. Greeves, M. S. Cherkiss, R. Hill, and F. J. Mazzotti. 2010. Thermoregulation by a brooding Burmese python (*Python molurus bivittatus*) in Florida. *Southeastern Naturalist* 9:403–405.

Van Wilgen, N. J., N. Roura-Pascual, and D. M. Richardson. 2009. A quantitative climate-match score for risk-assessment screening of reptile and amphibian introductions. *Environmental Management* 44:590–607.

Wall, F. 1912. A popular treatise on the common Indian snakes. *Journal of the Bombay Natural History Society* 21:447–476.

Walls, J. G. 1998. *The Living Pythons*. Neptune City, N.J.: T. F. H. Publications.

Willson, J. D., M. E. Dorcas, and R. W. Snow. 2011. Identifying plausible scenarios for the establishment of invasive Burmese pythons (*Python molurus*) in southern Florida. *Biological Invasions* DOI 10.1007/s10530-010-9908-3.

WEBSITES

Burmese Python: Species Profile. National Park Service. www.nps.gov/ever/naturescience/burmesepython.htm

Everglades Cooperative Invasive Species Management Area. www.evergladescisma.org

Florida Fish and Wildlife Conservation Commission Python Reporting Page. myfwc.com/contact/report/report-pythons/

Florida Invaders. www.nps.gov/ever/naturescience/floridainvaders.htm

Acknowledgments

Numerous people have devoted considerable time and effort to the invasive python issue, and we have worked with, collaborated with, or engaged in fruitful discussions of python biology, control methods, and directions for scientific research with many of them. Among the many volunteers, technicians, and researchers who have assisted in countless ways or provided essential data on pythons in South Florida we are especially grateful to Paul Andreadis, Kimberly Andrews, Rick Bauer, Matt Brien, Kristen Cecala, Mike Cherkiss, Justin Davis, Evan Eskew, Anthony Flanagan, Chris Gillette, Scott Goetz, Harry Greene, Wellington Guzman, Cris Hagen, Kristen Hart, Rebecca Harvey, Bobby Hill, Toren Hill, Josh Holbrook, Trey Kieckhefer, Kenneth Krysko, Frank Mazzotti, Melissa Miller, David Millican, Tony Mills, Lori Oberhofer, Shannon Pittman, Sean Poppy, Steven Price, Kenneth Rice, Michael Rochford, LeRoy Rodgers, Christina Romagosa, Ron Rozar, Dustin Smith, Charlotte Steelman, Theresa Walters, Lynea Witczak, and Alex Wolf. Lynea Witczak helped us organize the images for the book; Adrien Domske assisted with data management; and Whitney Webb gathered and organized much of the literature used in the text. Susan Harris and Jackie Guzy provided comments on the proofs, and Susan Harris wrote the index. Joy Vinci assisted with logistics and travel during our many trips to South Florida. Discussions with Mike Rochford, Chris Gillette, and Paul Andreadis provided useful information about various aspects of python biology.

We are grateful to the many people who graciously provided images that were used in this book. Their names are listed in the photo credits below.

Bob Reed graciously assisted us by answering numerous questions and discussing many of the issues and facts we present in this book. The exhaustive scientific report on the biology of large constrictors that Bob Reed and Gordon Rodda published in 2009 provided critical information important to developing many of the chapters in this book. We have worked with Kristen Hart on various python research projects and greatly appreciate her expertise, assistance, discussions, and encouragement. Paul Andreadis, Jeff Fobb, Chris Gillette, Kenneth Krysko, Bob Reed, and Skip Snow provided interesting anecdotes of their experiences with pythons. Frank Mazzotti is largely responsible for getting us involved in python research, and his support and encouragement have been critical over the last five years of collaborative research. Likewise, Mike Cherkiss has greatly assisted with various logistics related to our involvement in python research.

Judy Purdy, who is now retired from the University of Georgia Press, asked us to do this book, and her relentless enthusiasm and encouragement were inspirational to us. Bob Reed, Skip Snow, Steve Price, Kristen Hart, Whit Gibbons, and two anonymous reviewers all carefully read the manuscript, and their comments and suggestions greatly improved it.

Whit Gibbons has been a major source of professional and personal support for both of us for many years. His endless, youthful enthusiasm for ecology, herpetology, and conservation infects everyone he works with. Neither of us would be where we are today without the contributions Whit has made to our professional lives, and we certainly would not have had nearly as much fun along the way.

Skip Snow has been the central figure in invasive python research since it became an issue. His tireless work as the point person for all python reports and captures has been critical in providing the basic information that supports most of the research being conducted on invasive pythons. Likewise, his willingness to collaborate with numerous researchers has helped to provide a collegial atmosphere that fosters high-quality research. His unfailing graciousness, generosity with information, and logistical assistance have been critical to our work on various projects related to invasive pythons. This book, and in fact most of the research conducted on pythons, would not have been possible without Skip.

Finally, our families and loved ones were particularly understanding and helpful while we were working on this book. Michael E. Dorcas wishes to express his love and appreciation to his wife, Tammy, and children, Taylor, Jessika, and Zachary Dorcas. Their unending encouragement, understanding, and support are incalculable. John D. Willson wishes to thank Sarah DuRant and his family, Tom, Janet, and Susan Willson, for fostering his passion for reptiles and providing constant support and enthusiasm.

Partial support for our research was provided by the U.S. Geological Survey through the University of Florida, Davidson College, Duke Energy, the J. E. and Majorie Pittman Foundation, the Associated Colleges of the South, the Department of Energy under Award Number DE-FC-09-075R22506, and by a National Science Foundation grant to M. E. Dorcas.

Photo Credits

The authors thank the following individuals and organizations for providing photographs:

Graham Alexander Photograph on page 27 (bottom).

Paul Andreadis Photographs on pages 28, 45, and 79.

Kimberly Andrews Photograph on page 34 (top).

Mark Auten Photographs on pages 24 (bottom left and bottom right) and 51.

Michael Avery Photograph on page 65.

Brady Barr Photographs on pages 38 (bottom), 103, 138 (both), 139, and 141 (top).

Michael Barron (NPS photo) Photographs on pages 3 (left) and 84 (top).

Richard D. Bartlett Photographs on pages 8, 27 (top), 123 (all), and 142.

Jemeema Carrigan Photographs on pages 43 (bottom left) and 93.

Kristen Cecala Photograph on pages ii–iii.

Isaac Chellman Photograph on page 96 (bottom).

Crossroads Creative / istockphoto.com Base maps on pages 61, 117, and 133.

Clay DeGayner Photograph on page 99 (bottom right).

Bob DeGross (NPS photo) Photograph on page 4.

Lutz Dirksen Photographs on pages 136 and 137 (top).

Rakesh Kumar Dogra Photograph on page 35 (top).

Michael E. Dorcas Photographs on pages 5, 6 (middle), 15 (bottom), 16 (bottom), 52, 58 (top), 97, 101, and 126 (right).

Jeffrey Fobb Photographs on pages 36, 104, and 109.

Karen Garrod (USGS photo) Photographs on pages 46 (top left) and 88 (bottom).

Christopher Gillette Photographs on pages vi, viii, xii, 9, 14, 22, 56, 59 (left and top right), 60, 63, 82, 91, 98, 99 (top and bottom left), and 102.

Scott Goetz (USGS photo) Photographs on pages 113 (left) and 114.

Jo Greenhut Photograph on page 62.

Cris Hagen Photographs on pages 31 (both), 35 (bottom), 39, 87 (top), 120, and 121.

Robert Hill Photographs on pages 86 (right), 92, and 110 (right).

Brian D. Horne Photographs on pages 26, 38 (top), and 126 (left).

Ansar Khan Photograph on page 23.

Kenneth Krysko Photograph on page 73.

Bjorn Lardner Photograph on page 140.

Kris Leefers (NPS photo) Photograph on page 83 (top).

Janice Lynch (NPS photo) Photograph on page 111 (right).

Julie Larsen Maher © Wildlife Conservation Society Photographs on pages 29 and 125 (top).

Melissa Miller Photographs on pages 43 (top) and 128.

Tony Mills Photograph on page 16 (top).

Fred Minderman (NPS photo) Photograph on page 90.

Jody Mitchell (NPS photo) Photograph on page 86 (left).

Gary Nafis Photographs on pages 30 (top) and 33.

National Park Service (photographer unknown) Photograph on page 141 (bottom).

Lori Oberhofer (NPS photo) Photographs on pages v, 12 (bottom), 20, 34 (bottom), 41, 44 (both), 47 (top left and bottom), 58 (bottom), 70 (both), 83 (bottom left and bottom right), 84 (bottom), 87 (bottom), 89, 94 (right), 100, 110 (left), and 130.

Michelle Peake (NPS photo) Photograph on page 115.

Gad Perry Photograph on page 111 (left).

Todd Pierson Photographs on pages xi and 2.

Shannon Pittman Photographs on pages 46 (right), 50, 57, and 81.

Carol J. Plymale Photograph on page 3 (right).

Mike Praznovsky Jr. Photograph on page 30 (left).

Priyanka (Flickr name Priyasavy) Photographs on pages 30 (bottom right) and 37.

Steve Raymer / National Geographic Stock Photograph on page 124.

Robert Reed (USGS photo) Photographs on pages 42 (left), 67, 132 (right), and 135.

Michael Rochford Photographs on pages 1, 19, and 80.

Christina Romagosa Photograph on page 132 (left).

Ron Rozar (USGS photo) Photographs on pages 24 (top), 69 (right), 113 (right), 119, 134, and 143.

Sandra Rozar Photograph on page 42 (right).

Skip Snow (NPS photo) Photographs on pages 46 (bottom left), 88 (top), 95, 112, 131 (both), and 137 (bottom).

Chris Sondreal Photographs on pages i and 54.

Timothy Taylor Photograph on page 7 (left).

Marsha Ward Photograph on page 108 (top).

John D. Willson Photographs on pages 6 (left and right), 7 (right), 10 (all), 11 (both), 12 (top), 15 (top), 17, 43 (bottom right), 47 (top right), 53 (both), 59 (bottom right), 68, 69 (left), 71, 76, 78, 85, 96 (top), 99 (bottom middle), 106, 118 (both), and 125 (bottom).

Shona Wilson Photograph on page 94 (left).

Alex Wolf Photographs on pages 40, 48, 49, 107, and 108 (bottom).

Index

Italicized page numbers refer to illustrations.